MCDP 1-0

Marine Corps Operations

U.S. Marine Corps

PCN 142 000014 00

DEPARTMENT OF THE NAVY
Headquarters United States Marine Corps
Washington, DC 20308-1775

27 September 2001

FOREWORD

Marine Corps Doctrinal Publication (MCDP) 1-0 is the first Marine Corps Operations doctrinal publication written for the Marine component and the Marine air-ground task force (MAGTF). It represents how our warfighting philosophy is codified in operational terms. MCDP 1-0 is intentionally written broadly to capture the employment of Marine components and the MAGTF across the range of military operations. The supporting tactics, techniques, and procedures (TTP) are contained in our warfighting and reference publications.

This doctrinal publication is the transition—the bridge—between the Marine Corps' warfighting philosophy of maneuver warfare to the TTP used by Marines. MCDP 1-0 is written for the Marine component and the MAGTF and is the precursor to future MAGTF-oriented warfighting doctrine. It addresses how the Marine Corps conducts operations to support the national military strategy across the broad range of naval, joint, and multinational operations. MCDP 1-0 explores the contribution to the national defense provided by the unique structure of Marine Corps organizations—the Marine component and the MAGTF. It reflects the notion that "words matter", providing definitions of key operational terms to ensure that Marines speak a common operational language. It describes the role of the Marine component in providing, sustaining, and deploying Marine Corps forces at the operational level of war and how the MAGTF conducts expeditionary operations at the operational and tactical levels.

MCDP 1-0 focuses on how Marine Corps forces conduct operations today and the direction the Marine Corps capstone concept *expeditionary maneuver warfare* and the supporting concepts of *operational maneuver from the sea* and *ship-to-objective maneuver* will take Marine Corps operations in the near future.

MCDP 1-0 concentrates on the operating forces of the Marine Corps—Marine Corps forces as a Service component under joint force command, and the MAGTF. It acknowledges that Marine Corps operations are now and will continue to be joint and likely multinational. It describes how Marine Corps forces support the joint or multinational force commander and what capabilities the Marines bring to a joint or multinational force. It illustrates how the Marine Corps' task-organized combined arms forces, flexibility, and rapid deployment capabilities apply to the widening spectrum of employment of today's military forces. This publication provides the fundamentals of how MAGTFs conduct tactical operations and concisely addresses the types of operations MAGTFs will conduct to accomplish these missions. Marine Corps commanders and leaders at all levels should read and study this publication. Additionally, joint force commanders and their staffs can use this publication to better understand the capabilities of Marine Corps forces assigned to the joint force.

/s/

J. L. JONES
General, United States Marine Corps
Commandant of the Marine Corps

DISTRIBUTION: 142 000014 00

©2001 United States Government as represented by the Secretary of the Navy. All rights reserved.

Throughout this publication, masculine nouns and pronouns are used for the sake of simplicity. Except where otherwise noted, these nouns and pronouns apply to either gender.

MCDP 1-0, MARINE CORPS OPERATIONS

TABLE OF CONTENTS

Chapter 1 The Marine Corps in National Defense

Historical Role	1-3
The Character of Modern Conflict	1-6
Operational Environment	1-7
Threat Dimension	1-7
Political Dimension	1-9
Levels of War	1-9
Range of Military Operations	1-10
The National Security Structure	1-10
National Command Authorities	1-10
The Chairman of the Joint Chiefs of Staff	1-10
The Joint Chiefs of Staff	1-11
Unified Action	1-11
Joint Operations	1-11
Multinational Operations, Alliances, and Coalitions	1-12
Roles and Functions	1-13
Title 10, United States Code, *Armed Forces*	1-13
Goldwater-Nichols Department of Defense Reorganization Act of 1986	1-14
Department of Defense Directive 5100.1, *Functions of the Department of Defense and its Major Components*	1-15
Marine Corps Manual	1-16
Commandant of the Marine Corps	1-16
Organization and Structure	1-17
Headquarters, U.S. Marine Corps	1-17
Operating Forces	1-18
Supporting Establishment	1-22
Marine Corps Forces Reserve	1-22
Marine Corps Ethos	1-23

Chapter 2 Marine Corps Expeditionary Operations

Marine Corps Core Competencies	2-2
Warfighting Culture and Dynamic Decision-making	2-2
Expeditionary Forward Operations	2-2
Sustainable and Interoperable Littoral Power Projection	2-3
Combined Arms Integration	2-3
Forcible Entry from the Sea	2-3
Expeditionary Operations	2-4
Force Projection	2-5
Amphibious Operations	2-6
Amphibious Operation Command Relationships	2-6
Forcible Entry through an Amphibious Assault	2-8
Sequence	2-9
Maritime Pre-positioning Force	2-11
Sustained Operations Ashore	2-12
Enabling Force	2-12
Decisive Force	2-12
Exploitation Force	2-13
Sustaining Force	2-13
Emerging Concepts and Technologies	2-13
Expeditionary Maneuver Warfare	2-14
Operational Maneuver from the Sea	2-15
Ship-to-Objective Maneuver	2-16
Maritime Pre-positioning Force Future	2-17
Expeditionary Bases and Sites	2-18

Chapter 3 Marine Corps Forces

Marine Corps Component	3-1
Role and Responsibilities to the Commandant	3-3
Role and Responsibilities to a Combatant Commander	3-3
Role and Responsibilities to the Joint Force Commander	3-4
Joint Operations Conducted Through	
Service Component Commanders	3-5
Joint Operations Conducted Through	
Functional Component Commanders	3-5
The Marine Corps Component Commander as a	
Functional Component Commander	3-7

Role and Responsibilities to the MAGTF Commander.	3-9
Component Command Relationships and Staff Organization.	3-10
The Marine Air-Ground Task Force	3-10
Capabilities	3-11
Elements	3-12
Supporting Establishment	3-15
Types.	3-16
Marine Logistics Command.	3-20

Chapter 4 Employment of Marine Corps Forces at the Operational Level

Battlespace Organization	4-3
Area of Operations.	4-4
Area of Influence.	4-6
Area of Interest	4-7
Boundaries, Maneuver Control Measures, and Fire Support Coordinating Measures	4-7
Deployment	4-10
Force Deployment Planning and Execution	4-10
Predeployment Activities	4-14
Movement to and Activities at the Port of Embarkation.	4-15
Movement to the Port of Debarkation.	4-16
Reception, Staging, Onward Movement, and Integration	4-16
Deployment Forces	4-16
Employment.	4-18
Planning Approach	4-19
Arranging Operations	4-20
Combat Power.	4-21
Warfighting Functions.	4-21
Information Operations	4-22
Redeployment.	4-24
Expeditionary Operations in Support of Future Campaigns	4-24

Chapter 5 Logistics in Marine Corps Operations

Levels	5-3
Strategic Logistics.	5-4
Operational Logistics.	5-4
Tactical Logistics.	5-5

Functions	5-5
Supply	5-6
Maintenance	5-6
Transportation	5-7
General Engineering	5-7
Health Services	5-7
Services	5-7
Joint and Multinational Operations	5-8
Strategic Logistics Support	5-10
Strategic Mobility	5-10
United States Transportation Command	5-11
Department of Transportation	5-12
Defense Logistics Agency	5-12
Operational Logistics Support	5-12
Marine Corps Component	5-12
Marine Logistics Command	5-13
Reconstitution	5-14
Command and Control	5-15
Logistics Support to MAGTF Operations	5-16
Marine Expeditionary Force	5-17
Marine Expeditionary Brigade	5-18
Marine Expeditionary Unit (Special Operations Capable)	5-18
Special Purpose MAGTF	5-18
Air Contingency Force	5-19
Maritime Pre-positioning Forces	5-19
Aviation Logistics Support Ship	5-19
Norway Geopre-positioning Program	5-19
War Reserve Materiel Support	5-20
Tactical Level Command and Control	5-20
Supporting Establishment	5-20
Marine Corps Combat Development Command	5-21
Marine Corps Materiel Command	5-21
Marine Corps Bases, Stations, and Reserve Support Centers	5-22
Department of the Navy Agencies	5-22
Future Sustainment Operations	5-23

Chapter 6 Planning and Conducting Expeditionary Operations

Maneuver Warfare	6-2
Operational Design	6-3
Visualize	6-5
Describe	6-5
Direct	6-6
Planning	6-9
Commander's Battlespace Area Evaluation	6-10
Analyze and Determine the Battlespace	6-10
Centers of Gravity and Critical Vulnerabilities	6-11
Commander's Intent	6-12
Commander's Critical Information Requirements	6-13
Commander's Guidance	6-14
Mission	6-15
Decisive Action	6-17
Shaping Actions	6-18
Battlefield Framework	6-19
Integrated Planning	6-19
Single Battle	6-20
Deep Operations	6-21
Close Operations	6-22
Rear Operations	6-22
Noncontiguous and Contiguous	6-23
Main and Supporting Efforts	6-24
The Reserve	6-27
Security	6-28
Phasing	6-29
Operation Plans and Orders	6-30
Transitioning Between Planning and Execution	6-31
Execution	6-31
Command and Control	6-32
Assessment	6-33
Tactical Tenets	6-34
Achieving a Decision	6-35
Gaining Advantage	6-36
Tempo	6-38
Adapting	6-39
Exploiting Success and Finishing	6-40

Chapter 7 The MAGTF in the Offense

Purpose of Offensive Operations	7-2
Characteristics of Offensive Operations	7-3
Organization of the Battlespace	7-3
Organization of the Force	7-4
Types of Offensive Operations	7-7
Movement to Contact	7-7
Attack	7-10
Exploitation	7-14
Pursuit	7-15
Forms of Maneuver	7-16
Frontal Attack	7-17
Flanking Attack	7-18
Envelopment	7-19
Turning Movement	7-21
Infiltration	7-22
Penetration	7-23
Future Offensive Operations	7-24

Chapter 8 The MAGTF in the Defense

Purpose of Defensive Operations	8-3
Characteristics of MAGTF Defensive Operations	8-4
Preparation	8-7
Security	8-7
Disruption	8-7
Mass and Concentration	8-8
Flexibility	8-8
Maneuver	8-8
Operations in Depth	8-8
Organization of the Battlespace	8-10
Security Area	8-11
Main Battle Area	8-12
Rear Area	8-12
Organization of the Force	8-12
Security Forces	8-12
Main Battle Forces	8-13
Rear Area Forces	8-15

Types of Defensive Operations	8-16
Mobile Defense	8-17
Position Defense	8-20
Future Defensive Operations	8-22

Chapter 9 Other MAGTF Tactical Operations

Retrograde	9-1
Delay	9-2
Withdrawal	9-3
Retirement	9-4
Passage of Lines	9-5
Linkup	9-6
Relief in Place	9-7
Obstacle Crossing	9-7
Breach	9-8
River Crossing	9-9
Breakout from Encirclement	9-9

Chapter 10 Military Operations Other Than War

Principles	10-3
Objective	10-4
Unity of Effort	10-4
Security	10-5
Restraint	10-5
Perseverance	10-6
Legitimacy	10-6
Arms Control	10-7
Combatting Terrorism	10-7
Department of Defense Support to Counterdrug Operations	10-7
Enforcement of Sanctions/Maritime Intercept Operations	10-8
Enforcing Exclusion Zones	10-8
Ensuring Freedom of Navigation and Overflight	10-8
Humanitarian Assistance	10-8
Military Support to Civil Authorities	10-9
Nation Assistance/Support to Counterinsurgency	10-10
Noncombatant Evacuation Operations	10-11
Peace Operations	10-12
Protection of Shipping	10-13
Recovery Operations	10-13
Show of Force Operations	10-14

Strikes and Raids	10-14
Support to Insurgency	10-14
Warfighting Functions	10-14
Command and Control	10-15
Maneuver	10-16
Fires	10-16
Intelligence	10-17
Logistics	10-18
Force Protection	10-19

Chapter 11 MAGTF Reconnaissance and Security Operations

MAGTF Reconnaissance Assets	11-2
Command Element	11-3
Ground Combat Element	11-4
Aviation Combat Element	11-5
Combat Service Support Element	11-6
National and Theater Assets	11-6
Reconnaissance Planning	11-6
Types of Reconnaissance Missions	11-8
Route Reconnaissance	11-8
Area Reconnaissance	11-8
Zone Reconnaissance	11-9
Force-Oriented Reconnaissance	11-9
Reconnaissance Pull and Reconnaissance Push	11-9
Counterreconnaissance	11-11
Security Forces and Missions	11-12
Screen	11-13
Guard	11-14
Cover	11-16
Security Operations in Other Tactical Operations	11-17
Military Operations Other Than War	11-17

Appendices

A Warfighting Functions	A-1
B Principles of War	B-1
C Tactical Tasks	C-1
D Planning and Employment Considerations for Tactical Operations	D-1
E Planning and Employment Considerations for MOOTW	E-1
F Glossary	F-1

Bibliography

Bibliography-1

Figures

1-1. Marine Corps Organization	1-18
1-2. Marine Corps Forces and Fleet Marine Force Relationship	1-21
2-1. Current Location of Maritime Pre-positioning Ships Squadrons	2-11
2-2. Operational Objectives	2-17
3-1. Chains of Command and Command Relationships	3-2
3-2. Commander's Level of War Orientation	3-4
3-3. Combatant Command Organized by Functional Components	3-6
3-4. Marine Air-Ground Task Forces	3-12
3-5. MAGTF Organization	3-13
4-1. Noncontiguous Areas of Interest	4-8
4-2. Unit Boundaries	4-9
5-1. Strategic Mobility Considerations	5-11
6-1. Operational Design	6-4
6-2. Commander's Vision of Decisive and Shaping Actions and Sustainment	6-7
6-3. Battlefield Framework	6-8
6-4. Single Battle	6-21
6-5. Battlespace Organization	6-24
7-1. Types of Offensive Operations	7-7
7-2. Movement to Contact	7-8
7-3. Pursuit	7-16
7-4. Frontal Attack	7-17
7-5. Flanking Attack	7-18
7-6. Single Envelopment	7-19
7-7. Double Envelopment	7-20
7-8. Turning Movement	7-21
7-9. Infiltration	7-22
7-10. Penetration	7-24
8-1. Organization of the Battlespace	8-11
8-2. Organization of the Force	8-13
8-3. Mobile Defense	8-18
8-4. Position Defense	8-21
9-1. Delay	9-3

CHAPTER 1

The Marine Corps in National Defense

Contents	
Historical Role	**1-3**
The Character of Modern Conflict	**1-6**
Operational Environment	**1-7**
Threat Dimension	1-7
Political Dimension	1-9
Levels of War	1-9
Range of Military Operations	1-10
The National Security Structure	**1-10**
National Command Authorities	1-10
The Chairman of the Joint Chiefs of Staff	1-10
The Joint Chiefs of Staff	1-11
Unified Action	**1-11**
Joint Operations	1-11
Multinational Operations, Alliances, and Coalitions	1-12
Roles and Functions	**1-13**
Title 10, United States Code, *Armed Forces*	1-13
Goldwater-Nichols Department of Defense Reorganization Act of 1986	1-14
Department of Defense Directive 5100.1, *Functions of the Department of Defense and its Major Components*	1-15
Marine Corps Manual	1-16
Commandant of the Marine Corps	**1-16**
Organization and Structure	**1-17**
Headquarters, U.S. Marine Corps	1-17
Operating Forces	1-18
Supporting Establishment	1-22
Marine Corps Forces Reserve	1-22
Marine Corps Ethos	**1-23**

"Despite its outstanding record as a combat force in the past war [World War II], the Marine Corps' far greater contribution to victory was doctrinal [new concepts]: that is, the fact that the basic amphibious doctrines which carried Allied troops over every beachhead of World War II had been largely shaped—often in [the] face of uninterested or doubting military orthodoxy—by U.S. Marines, and mainly between 1922 and 1935."
—General Alexander A. Vandegrift, USMC

"...American history, recent as well as remote, has fully demonstrated the vital need for the existence of a strong force-in-readiness. Such a force, versatile, fast moving, and hard-hitting,... can prevent the growth of potentially large conflagrations by prompt and vigorous action during their incipient stages. The nation's shock troops must be the most ready when the nation is least ready... to provide a balanced force-in-readiness for a naval campaign and, at the same time, a ground and air striking force ready to suppress or contain international disturbances short of large scale war...."
-82nd Congress (1952)

Since 1775, the United States Marine Corps has served as an expeditionary force organized and trained to act in the national security interest and carry out the national military strategy. The Marine Corps' contribution to the national defense has successfully evolved throughout its history by virtue of the ability of Marines to identify and adapt to the nation's national security needs, often before those needs were commonly recognized. Such innovations as the seizure and defense of advanced Naval bases, amphibious operations, close air support, helicopterborne vertical envelopment tactics, maritime pre-positioning forces (MPFs), and task-organized, combined arms forces consisting of aviation, ground, and logistic elements known as Marine air-ground task forces (MAGTFs) are prime examples of how the Marine Corps has adapted and evolved as an expeditionary force. The Marine Corps continually reviews its roles and missions in the context of an uncertain world, adapting to the changing security needs of the Nation while preserving those core values and professional capabilities that make Marines succeed in war and peace.

The Nation requires an expeditionary force-in-readiness that can respond to a crisis anywhere in the world. The Marine Corps provides self-sustainable, task organized combined arms forces capable of conducting a full spectrum of operations in support of the joint force commander. Missions might include forcible entry operations, peace enforcement, evacuation of American citizens and embassies, humanitarian assistance or operations to reinforce or complement the capabilities of other Services to provide balanced military forces to the joint force commander. The unique capability of the Marine Corps as a sea service and partner with the United States Navy allows the use of the sea as both a maneuver space and a secure base of operations to conduct operations in the littoral areas of the world. The ability to remain at sea for long periods without the requirement of third nation basing rights makes the Marine Corps the force of choice in emerging crises. Marine Corps forces exploit the Total Force concept, employing combinations of active duty and reserve Marines to ensure that missions are effectively and efficiently executed.

Naval expeditions comprised of Navy and Marine Corps forces have long been the instruments of choice in our Nation's response to global contingencies. From humanitarian assistance, to peacekeeping, to combat, these forces are normally the first on scene and ready to respond. Naval expeditionary forces combine the *complementary but distinct capabilities* of the Navy and Marine Corps. They provide strategic agility and overseas presence without infringing on the sovereignty of other nations and simultaneously enable enhanced force protection. They provide a power projection capability that can be tailored to meet a wide range of crises from a major theater war to military operations other than war (MOOTW). Naval expeditionary forces can be task-organized to provide an array of options to the National Command Authorities and combatant commanders in dealing with a particular situation. Naval expeditionary forces provide the United States the unique capability to conduct and sustain operations from the sea—including continuous forward presence and self-sustainment—in support of our national interests without reliance on pre-positioning ashore, foreign basing or the granting of overflight rights.

HISTORICAL ROLE

Throughout its history, the Marine Corps has lead in developing innovative and successful military concepts. These concepts have helped the Marines and their sister Services to win the Nation's battles and wars. The Marine Corps success in developing into the Nation's premier expeditionary force has its roots in decades of innovative thought, bold experiments, and constant training.

The Marine Corps was established in 1775 to provide landing forces (LFs) for Navy ships. Throughout its first 150 years of existence the Marine Corps provided Marines for ships detachments and temporary battalions and brigades formed from ships detachments and Marine Barracks to provide LFs to the fleet during naval expeditions. These forces conducted expeditionary operations throughout the world such as—

- Seizing New Providence in the Bahamas from the British (1776).
- Raising the United States flag over a foreign city for the first time when Lieutenant O'Bannon and eight Marines, leading a rag-tag force of Arabs and mercenaries, captured the Tripolitan city of Derna (1805).
- Accompanying Commodore Perry as he opened Japan to trade with the world (1854).

- Conducting an amphibious assault on Fort Fisher in the Civil War (1865).
- Landing at Guantanamo Bay during the Spanish American War (1898).

Marines participated with Army forces in sustained combat operations as in the capture of Mexico City (1847) and in the Meuse-Argonne offensive during World War I (1918). Marines also conducted other special missions as directed by the President, such as the capture of John Brown at Harper's Ferry (1859) and the defense of the American Legation in Peking during the Boxer Rebellion (1900). Deployed in MOOTW, Marines performed constabulary and nation building duties in the Philippines, Cuba, Nicaragua, and Haiti (1899-1934).

In the years between World War I and World War II, the Marines, building on their long experience as an expeditionary force from the sea, created the concept of modern amphibious warfare. The Marines studied past operations, experimented with new equipment such as landing craft and amphibious assault vehicles, and conducted innovative amphibious exercises with the fleet. These efforts resulted in the doctrine and new tactics, techniques, and procedures in amphibious warfare that were instrumental in winning in the European and Pacific theaters of war during World War II. Not only was this new concept of amphibious warfare displayed at Tarawa, Saipan, and Iwo Jima, the landings at North Africa, Sicily, Salerno, and Normandy were also a direct result of the conceptual effort of the Marines. Lessons learned from World War II operations were applied in the masterful Inchon landing during the Korean War (1950), and the continued evolution of amphibious operations enabled Marine Corps expeditionary operations through the end of the 20th century.

The Marine Corps concept of creating expeditionary combined arms forces that exploited the synergy of task-organized Marine aviation, ground combat forces and combat service support, was codified by the National Security Act of 1947. The MAGTF concept was soon tested in combat with the rapid deployment and highly successful operations of the 1st Marine Brigade (Provisional) in the early days of the Korean War. This MAGTF was rapidly formed and deployed and was instrumental in stopping the North Korean's offensive to drive United States forces from Korea. The 1st Marine Brigade's defense of the Pusan Perimeter enabled the decisive Inchon landing that led to the defeat and expulsion of the North Koreans. Public Law 416 passed by the 82nd Congress in 1952 further solidified the nature of the MAGTF. This law ensured that the Marine Corps would be organized with three aircraft wings and three combat divisions.

Throughout the 1940s and 1950s, the Marines led the way in developing concepts of employment for the helicopter, using this new technology to rapidly move troops, supplies, and evacuate casualties during the Korean War. The concept of vertical envelopment added a new dimension not only to the amphibious assault but to a range of military operations soon to be conducted in Algeria (1954-62), the Sinai (1956), and by Marines and the Army in Southeast Asia (1962-75).

Building on the counterinsurgency experiences of Marines in Haiti and Nicaragua, innovative Marines created the combined action platoon (CAP) program in South Vietnam in 1965. This program placed small teams of Marines, led by noncommissioned officers, in the hamlets and villages throughout the Marines' area of operations (AO). These Marines earned the trust of the villagers by living in the village while protecting the people. Marines led and trained the local people's defense forces, learned the language and customs of the villagers, and were very successful in denying those areas under their control to the enemy. The CAP program became a model for success in countering insurgencies. Many of the lessons learned from the CAP program were emulated in various peace enforcement and humanitarian assistance operations Marines have performed over the last decade, such as Operation Provide Comfort in Northern Iraq (1991) and Operation Restore Hope in Somalia (1992-93).

Following the Vietnam War and military operations such as the ill-fated Iranian rescue mission, Operation Urgent Fury in Grenada, and peace enforcement in Beirut in the 1970s and early 1980s, Congress mandated several important changes in the way that joint forces conducted military operations. The Goldwater-Nichols Department of Defense Reorganization Act of 1986 defined the responsibilities of the Chairman of the Joint Chiefs of Staff, Service chiefs, and combatant commanders and provided the authority to perform them. The law fundamentally changed the way the Department of Defense operated by distinguishing between the operational contributions of the Services and those of the combatant commands. At the operational and tactical levels, the Services were to provide forces to the joint force commander who would then exercise command in operations. This combination of air, ground, and sea forces closely mirrored the organization of the MAGTF. Accordingly, Marines experienced no difficulty in understanding and adapting to joint operations. After all, the MAGTF was the precursor to the functionally oriented joint force. The joint operations concept was validated in Operation Just Cause in Panama (1989) and Operation Desert Shield/Desert Storm in the Persian Gulf (1990-91).

The inability of the Nation to rapidly deploy large numbers of heavy forces to the Middle East in the 1970s and 1980s provided the impetus to the Marine Corps to develop and refine the MPF concept. The Marines helped the Navy design the ships of the maritime pre-positioning ships squadrons (MPSRON), identifying the key requirements and features necessary to provide rapid loading and unloading, adequate storage, and supply capacity. The Marines developed and tested the organization of the squadrons and those Marine units that would execute the loading, maintenance, and off loading of the maritime pre-positioning ships (MPS). The MPFs concept has been validated with the extremely successful employment of MPF in combat operations during Operation Desert Shield/Desert Storm, and in humanitarian assistance operations such as Operation Sea Angel in Bangladesh (1991) and Operation Restore Hope in Somalia (1992-93). This concept has provided the combatant commanders with a very flexible capability to rapidly deploy and sustain robust combat and humanitarian forces throughout the littorals.

Maneuver warfare is the Marine Corps warfighting philosophy and forms the basis for the concept of expeditionary maneuver warfare. During the late 1970s and the 1980s Marines embraced the theory of maneuver warfare and developed their own institutional approach to maneuver warfare. This process of debate, discussion, and experimentation culminated in the publication of Fleet Marine Force Manual 1, *Warfighting*. This seminal document subsequently provided the foundation for the training and education of Marine leaders who conducted maneuver warfare with great success in Operation Desert Storm. *Warfighting* was followed by a series of doctrinal publications that provided further guidance on the theory and nature of strategy, campaigning, and tactics in maneuver warfare.

The continued development of new concepts and doctrine, along with the refining of accepted doctrine, will help ensure that the Marine Corps provides the Nation with a balanced force in readiness to conduct expeditionary operations in a dangerous and uncertain world.

THE CHARACTER OF MODERN CONFLICT

Despite (or perhaps as a result of) the victory of the United States in the Cold War, the world is characterized by crisis and disorder. Instead of the relative certainty provided by the bi-polar world of the United States and the Union of Soviet Socialist Republics, the world now has a single super power—the United States—and several regional powers. This has led to regional competition in many areas of the world and new threats of terrorism and international organized crime have taken on more menacing dimensions. Burgeoning populations in emerging nations are threatened by famine, shortages of natural resources, and disease, and are particularly vulnerable to natural disasters. All of these factors combine to create an environment where United States military forces are constantly challenged to accomplish multidimensional missions. Marines may be conducting combat, peace enforcement, and humanitarian assistance operations simultaneously within an emerging nation in an austere theater or a major metropolitan city in the littorals.

OPERATIONAL ENVIRONMENT

The environment where Marines will operate in the next decade reflects the changing character of modern conflict. While the Marine Corps has participated in some combat in urban areas such as Inchon and Seoul in the Korean War, Santo Domingo in the Dominican Republic (1965), Hue in the Vietnam War (1968), and Mogadishu in Somalia (1992-93), most Marine Corps combat operations have taken place in isolated unpopulated settings. The growth of cities along the world's littorals means Marines must be prepared to conduct a range of operations in heavily populated urban areas and that civilians and other noncombatants will be an ever-increasing concern to the commander. While fighting and winning the Nation's battles will ever be the focus of the Marine Corps, commanders must be able to successfully conduct a wide variety of missions against diverse and ever-changing threats while under a myriad of political constraints and restraints and while operating under joint or multinational command.

Threat Dimension

As the nature of modern conflict evolves so does the nature of the threat that the Nation and Marine Corps forces face. No longer are Marines only concerned with defeating Soviet or their proxy forces. Marines must be able to face threats that include natural or man-made disasters such as famine or disease, transnational criminals such as drug cartels and weapon smugglers, terrorists, and insurgents, as well as soldiers in the armies of the Nation's foes.

The national armed forces of regional powers continue to be a major threat to world peace. Regional powers and smaller nations continue to field large and relatively competent forces. These forces will continue to modernize their equipment to take advantage of technological advancements and to challenge United States strengths in information technology, intelligence gathering and targeting, and precision strike capabilities. United States forces must remain focused on the challenge of defeating these near peer competitors and regional powers to protect the Nation's interests.

Nonstate actors are those individuals or forces that perform criminal acts of terrorism, drug production and smuggling or conduct guerrilla warfare for personal gain or to achieve the nonlegitimate goals of their organization or cause. In some cases nations support or hire these nonstate actors to achieve their national goals.

Transnational entities conduct a wide variety of activities to threaten United States interests and security abroad and at home. These include terrorism, drug smuggling and production, illegal arms trading, organized international crime, and environmental terrorism. Disasters such as famine and disease, coupled with war and ethnic strife, result in migrations of refugees that overwhelm the resources of the gaining countries. International assistance is often the only recourse to alleviate the suffering of the refugees and prevent the gaining countries from closing their borders.

To succeed against United States forces, the enemy must be smart and flexible. Adversaries will develop forces, tactics, and equipment that will seek to attack perceived United States weaknesses and avoid strengths. These asymmetrical operations will require forces that can change tactics and organizations in anticipation of or in response to a new situation. For example, such an enemy force would conduct conventional force operations during early stages of a crisis while United States forces are deploying and the enemy has temporary superiority of forces. Once the United States has achieved conventional superiority in the AO, the enemy would then quickly adapt and conduct nonconventional operations, such as guerrilla warfare, terrorist attacks, and psychological operations.

Adversaries will conduct operations to set the conditions that lead to accomplishing their national or organizational objectives. They will try to change the nature of the conflict or employ capabilities that they believe the United States will have difficulties countering. Adversaries will attempt to shape the situation by—

- Conducting force-oriented operations against United States forces and coalition partners to create casualties.
- Denying or disrupting United States forcible entry operations.
- Avoiding decisive battles to preserve their forces and prolong the conflict.
- Conducting asymmetrical operations.
- Forming political and military coalitions to limit United States operations.

Adversaries will attempt to prolong the conflict while using terrorism and other attacks to erode international support, alliance cohesion, and the will to fight once United States forces arrive in the crisis area. Use of the media and other information operations to portray United States forces as aggressors, the use of noncombatants to garner world sympathy, and the fear generated by terrorism or the threat of employment of weapons of mass destruction are only some of the ways that an intelligent enemy can shape the situation. By turning world and United States' public opinion against the policies and practices of an alliance or international organization such as the North Atlantic Treaty Organization or the United Nations, the enemy can neutralize or eliminate United States military involvement in a crisis.

Political Dimension

The President identifies the United States security objectives in the *National Security Strategy*, establishing the Nation's broad political, economic, and military interests in peace and war. The *National Military Strategy* supports the national security strategy by translating strategic political and economic objectives into national military objectives and tasks. These objectives and tasks protect and promote United States national interests. Current national military objectives are to *promote peace and stability* and, when necessary, to *defeat adversaries* who threaten the United States, national interests or allies. These objectives drive strategic tasks: *shape* the international environment and create conditions favorable to United States interests and global security; *respond* to the full spectrum of crises; and *prepare* now for an uncertain future.

United States military forces accomplish the national military objectives and tasks through the four strategic concepts of *strategic agility*, *overseas presence*, *power projection*, and *decisive force*. The Nation's forward-deployed naval forces, together with forward and United States-based forces from all the Services, combine to meet the requirements of conflicts and can operate across the full range of military operations.

Levels of War

The highest level is the strategic. Strategy involves establishing goals, assigning forces, providing assets, and imposing conditions on the use of force. Strategy derives from political and policy objectives and is the sole authoritative basis for military operations. The strategic level of war involves the art of winning wars and maintaining the peace.

The next level—operational—links the strategic and tactical levels. It includes deciding when, where, and under what conditions to engage the enemy in battle. The operational level of war is the art and science of winning campaigns.

The final level of war is the tactical. Tactics are the concepts and methods used to accomplish a particular mission in either combat or MOOTW. In war, tactics focus on applying combat power to defeat an enemy force. The tactical level of war involves the art and science of winning engagements and battles to achieve the objectives of the campaign.

The distinctions between the levels of war are rarely clear and often overlap in practice. Commanders may operate at multiple levels simultaneously. In MOOTW, small unit leaders may conduct tactical actions that have operational and even strategic consequences.

Range of Military Operations

Conflict can take a variety of forms ranging from general war, such as a global conflict between major powers, all the way down to MOOTW where violence is limited and combat forces may not be needed. This range may be characterized by two major categories:

A *major theater war* is the employment of large joint and multinational forces in combat operations to defeat an enemy nation, coalition or alliance. Operation Desert Storm is an example of a major theater war.

A *smaller-scale contingency* normally encompasses a wide range of naval, joint or multinational operations in small wars and MOOTW. Peace enforcement operations in the Balkans and humanitarian assistance operations are examples of smaller-scale contingencies.

THE NATIONAL SECURITY STRUCTURE

The Marine Corps' involvement in national security has its foundation in law. The National Security Act of 1947 unified the defense establishment, assigned roles, missions, and functions among major Department of Defense agencies, including the Joint Chiefs of Staff, Military Services, and combatant commands.

National Command Authorities

The National Command Authorities include the President and the Secretary of Defense or their duly deputized alternates or successors. They exercise authority over the Armed Forces through combatant commanders and the Secretaries of the Military Departments and the chiefs of the Services for those forces not assigned to the combatant commanders. The National Command Authorities translate policy into national strategic military objectives.

The Chairman of the Joint Chiefs of Staff

In accordance with Title 10, United States Code, Sections 151 and 153, the Chairman of the Joint Chiefs of Staff assists the President and the Secretary of Defense in providing strategic direction of the Armed Forces. His responsibilities include—

- Presiding over the Joint Chiefs of Staff.
- Acting as the spokesman for the combatant commanders.
- Preparing military strategy, assessments, and strategic plans.
- Providing for the preparation and review of joint operation plans.
- Providing military guidance to the Services in preparation of their detailed plans.

The Joint Chiefs of Staff

The Joint Chiefs of Staff consists of the Chairman of the Joint Chiefs of Staff, the Vice Chairman, and the four Service chiefs. Joint Chiefs of Staff members advise the National Command Authorities upon their request or when a member recognizes a matter of national security that merits being addressed. When the Chairman advises the President, National Security Council or the Secretary of Defense, any member may submit advice or an opinion in disagreement with that of the Chairman in addition to the advice provided by the Chairman. Duties of the members take precedence over all their other duties as Service chiefs.

UNIFIED ACTION

Unified action is a generic term that refers to a broad scope of activities (including the synchronization of activities with governmental and nongovernmental agencies) taking place within unified commands, sub-unified commands, or joint task forces under the overall direction of the commanders of those commands. The national military strategy calls for the Marine Corps to act as part of fully interoperable and integrated joint forces. The joint force commander synchronizes the employment of Marine Corps forces with that of the other Services to fully exploit the capabilities of the joint force and to effectively and efficiently accomplish the mission.

Joint Operations

Joint operations are operations that include forces of two or more Military Departments under a single commander. Joint force commanders use joint forces within their AOs to participate in engagement activities and to conduct military operations in support of the geographic combatant commanders' contingency and war plans. Combatant commanders and their staffs are responsible for preparing plans for engagement with other nations and their forces throughout the theater. They also must prepare and maintain contingency and war plans for their theater of operations.

Engagement is the use of military forces to contribute to regional stability, reduction of potential conflicts, and the deterrence of aggression. Engagement activities are pro-active, conducted at home and abroad, that take advantage of opportunities to shape the international security environment. They include overseas presence, forward deployment, foreign internal development, and alliance and coalition training and exercises.

Contingency plans are plans for major contingencies that can reasonably be expected to occur within the theater of operations. Contingency planning can be deliberate or can be conducted under time constraints in crisis action planning. Contingency plans differ from operation plans as they are in an abbreviated format and require significant time or considerable expansion or alteration to convert them to operation plans or orders. War plans are completed operation plans for major contingencies such as major theater wars.

One of the techniques the combatant commanders and their staffs employ in contingency and war planning is flexible deterrent options. These are tailored military forces and operations designed to effectively and efficiently accomplish an anticipated mission or meet an unexpected contingency. Joint planners use flexible deterrent options within a planning framework intended to facilitate early decision and rapid response by laying out a range of forces and capabilities to be employed by the combatant commanders to accomplish particular missions. Flexible deterrent options are carefully tailored to send the right signal to the threat and the United States and world public. These options should include the minimum military force necessary to accomplish the objectives and the coordinated use of economic, diplomatic, and political actions appropriate to the particular situation. For more on Marine Corps participation in joint operations see chapter 3.

Multinational Operations, Alliances, and Coalitions

Although the United States may act unilaterally when the situation requires, it pursues its national interests through alliances and coalitions when possible. Alliances and coalitions can provide larger and more capable forces, share the costs of the operation, and enhance the legitimacy of the operation in world and United States public opinion. Multinational operations are usually conducted within the structure of an alliance or coalition. Alliances normally have established agreements for long term objectives, developed contingency plans, and standardized some equipment and procedures to ease interoperability. Coalitions are normally established for shorter periods or for specific multinational operations. They normally do not have established procedures or standardized equipment.

However organized, multinational operations normally involve complex cultural issues, interoperability challenges, conflicting national command and control procedures, intelligence sharing, and other support problems. Even long established alliances experience some degree of these obstacles. Unity of command is difficult to achieve in multinational operations. To compensate for this, commanders concentrate on obtaining unity of effort between the participating national forces. Consensus building is the key element in building unity of effort in multinational operations. Multinational operations command and control is usually based on parallel or lead nation command and control structures. Parallel command requires coordinated political and senior military leadership to make decisions and transmit their decisions through existing chains of command to their deployed forces. This is the simplest to establish but limits tempo. Lead nation command and control requires that one nation (usually the one providing the preponderance of forces or capabilities) provides the multinational force commander and uses that nation's command and control system. Other nations' forces are then assigned as subordinate forces. Normally, this structure requires some integration of national staffs.

Multinational commanders must be prepared to accommodate differences in operational and tactical capabilities by nations within the combined force. The commander's intentions, clear guidance, and plans must be articulated to avoid confusion that might occur due to differences in doctrine and terminology. Detailed planning, wargaming, exchange of standing operating procedures and liaison officers, and rehearsals help to overcome procedural difficulties between nations. Finally, the commander should ensure that the missions assigned to nations within the multinational force reflect the specific capabilities and limitations of each national contingent. Mission success should not be jeopardized because of unrealistic expectations of the capabilities or political will of member forces.

ROLES AND FUNCTIONS

Roles are the broad and enduring purposes for which the Services and USSOCOM were established by Congress by law. Missions are the tasks assigned by the President or Secretary of Defense to the combatant commanders. Functions are specific responsibilities assigned by the President and Secretary of Defense to enable the Services to fulfill their legally established roles. Various laws, directives, and manuals establish the roles and functions of the Marine Corps and describe the general composition and responsibilities of the Marine Corps. The key sources are Title 10, United States Code, *Armed Forces*; Goldwater-Nichols Department of Defense Reorganization Act of 1986; Department of Defense Directive 5100.1, *Functions of the Department of Defense and Its Major Components;* and the Marine Corps Manual.

Title 10, United States Code, *Armed Forces*

Chapter 507, Section 5063 details the Marine Corps' composition and functions. The Marine Corps—
- Shall be organized to include not less than three combat divisions and three aircraft wings, and other organic land combat forces, aviation, and services.

- Shall be organized, trained, and equipped to provide Fleet Marine Forces of combined arms, together with supporting aviation forces, for service with the fleet in the seizure and defense of advanced naval bases and for the conduct of such land operations as may be essential to the prosecution of a naval campaign.

- Shall provide detachments and organizations for service on armed vessels of the Navy, shall provide security detachments for the protection of naval property at naval stations and bases, and shall perform such other duties as the President may direct. These additional duties may not detract from or interfere with the operations for which the Marine Corps is primarily organized.

- Shall develop, in coordination with the Army and Air Force, those phases of amphibious operations that pertain to the tactics, techniques, and equipment used by landing forces.

- Is responsible, in accordance with integrated joint mobilization plans, for the expansion of the peacetime components of the Marine Corps to meet the needs of war.

Goldwater-Nichols Department of Defense Reorganization Act of 1986

Salient features of the act are the—

- Service chiefs (Chief of Staff of the Army, Chief of Naval Operations, Chief of Staff of the Air Force, and Commandant of the Marine Corps) are responsible for organizing, training, and equipping Service forces, while combatant commanders are responsible for the planning and execution of joint operations.

- Chairman of the Joint Chiefs of Staff is the *principal* military advisor to the President, National Security Council, and the Secretary of Defense. While he outranks all other officers of the Armed Forces, he does not exercise military command over the combatant commanders, Joint Chiefs of Staff, or any of the Armed Forces.

- Joint Staff is under the exclusive direction of the Chairman of the Joint Chiefs of Staff. It is organized along conventional staff lines to support the Chairman and the other members of the Joint Chiefs of Staff in performing their duties. The Joint Staff does not function as an overall Armed Forces General Staff and has no executive authority.

- Operational chain of command is clearly established from the President through the Secretary of Defense to the combatant commanders.

Department of Defense Directive 5100.1, *Functions of the Department of Defense and its Major Components*

This directive defines the primary functions of the Marine Corps. Among these primary functions are to—

- Organize, train, equip, and provide Marine Corps forces to conduct prompt and sustained combat operations at sea, including sea-based and land-based aviation. These forces will seek out and destroy enemy naval forces, suppress enemy sea commerce, gain and maintain general naval supremacy, control vital sea areas, protect vital sea lines of communications, establish and maintain local superiority in an area of naval operations, seize and defend advanced naval bases, and conduct land, air, and space operations essential to a naval campaign.

- Provide Marine Corps forces of combined arms for service with the Navy to seize and defend advanced naval bases and to conduct land operations necessary for a naval campaign. In addition, the Marine Corps shall provide detachments and organizations for service on armed vessels of the Navy and provide security detachments for naval stations and bases.

- Organize, equip, and provide Marine Corps forces to conduct joint amphibious operations. The Marine Corps is responsible for the amphibious training of all forces assigned to joint amphibious operations.

- Organize, train, equip, and provide forces for reconnaissance, antisubmarine warfare, protection of shipping, aerial refueling, and minelaying operations.

- Organize, train, equip, and provide forces for air and missile defense and space control operations.

- Provide equipment, forces, procedures, and doctrine to conduct and support electronic warfare.

- Organize, train, equip, and provide forces to conduct and support special operations.

- Organize, train, equip, and provide forces to conduct and support psychological operations.

Functions to be accomplished together with other Services include develop—

- The doctrine, procedures, and equipment of naval forces for amphibious operations and the doctrine and procedures for joint amphibious operations.

- The doctrine, tactics, techniques, and equipment employed by landing forces in amphibious operations. The Marine Corps has primary responsibility for the development of landing force doctrine, tactics, techniques, and equipment that are of common interest to the Army and the Marine Corps.

- Doctrine, procedures, and equipment of interest to the Marine Corps for airborne operations not provided for by the Army.
- Doctrine, procedures, and equipment employed by Marine Corps forces in the conduct of space operations.

In addition to the above functions, the Marine Corps will perform such other duties as the President or the Secretary of Defense may direct. However, these additional duties must not detract from or interfere with the operations which the Marine Corps is primarily organized. These functions do not contemplate the creation of a second land army. Finally, the directive describes collateral functions of the Marine Corps to train its forces to—

- Interdict enemy land and air forces and communications through operations at sea.
- Conduct close air and naval support for land operations.
- Furnish aerial photography for cartographic purposes.
- Participate in the overall air effort, when directed.
- Establish military government, as directed, pending transfer of this responsibility to other authority.

Marine Corps Manual

The Marine Corps Manual adds three more functions. The Marine Corps shall—

- Maintain a Marine Corps Forces Reserve for the purpose of providing trained units and qualified individuals to be available for active duty in the Marine Corps in time of war or national emergency and at such other times as the national security may require.
- Provide Marine Corps officer and enlisted personnel in support of the Department of State security program overseas.
- Organize Marine Corps aviation, as a collateral function, to participate as an integral component of naval aviation in the execution of such other Navy functions as the fleet commanders may direct.

COMMANDANT OF THE MARINE CORPS

The Commandant has two vital functions—as a member of the Joint Chiefs of Staff and as Marine Corps Service Chief. His duties as a member of the Joint Chiefs of Staff take precedence over all other duties.

As a Joint Chiefs of Staff member, the Commandant may submit his advice or opinion to the Chairman of the Joint Chiefs of Staff when it is in disagreement with or provides additional insight to the Chairman of the Joint Chiefs of Staff's point of view. When the Chairman of the Joint Chiefs of Staff submits his advice or opinion to the President, the National Security Council, and the Secretary of Defense, the Chairman of the Joint Chiefs of Staff is obligated to submit any additional input from the Commandant or other members of the Joint Chiefs of Staff. When the Commandant is acting in his capacity as a military adviser, he may provide advice to the President, the National Security Council, and the Secretary of Defense when his opinion is requested.

The Commandant may make recommendations to Congress relating to the Department of Defense, providing he has informed the Secretary of Defense prior to the meeting with Congress. The Commandant will attend the regularly scheduled meetings of the Joint Chiefs of Staff. As long as his independence as a member of the Joint Chiefs of Staff is not impaired, the Commandant will keep the Secretary of the Navy informed of military advice given by the other members of the Joint Chiefs of Staff on matters that impact the Department of the Navy.

As the Service chief, the Commandant is subject to the authority, direction, and control of the Secretary of the Navy. He is directly responsible for the administration, discipline, internal organization, training, requirements, efficiency, and readiness of the Marine Corps. He is also responsible for the Marine Corps' materiel support system and accountable for the total performance of the Marine Corps.

ORGANIZATION AND STRUCTURE

The Marine Corps' organization consists of Headquarters, U.S. Marine Corps; the operating forces; the supporting establishment; and the Marine Corps Forces Reserve. See figure 1-1 on page 1-18.

Headquarters, U.S. Marine Corps

The Commandant presides over the daily activities of Headquarters, U.S. Marine Corps. Headquarters, U.S. Marine Corps provides staff assistance to the Commandant by—

- Preparing the Marine Corps for employment. This is accomplished through recruiting, organizing, supplying, equipping (including research and development), training, servicing, mobilizing, demobilizing, administering, and maintaining the Marine Corps.
- Investigating and reporting on the efficiency of the Marine Corps and its preparation to support military operations by combatant commanders.

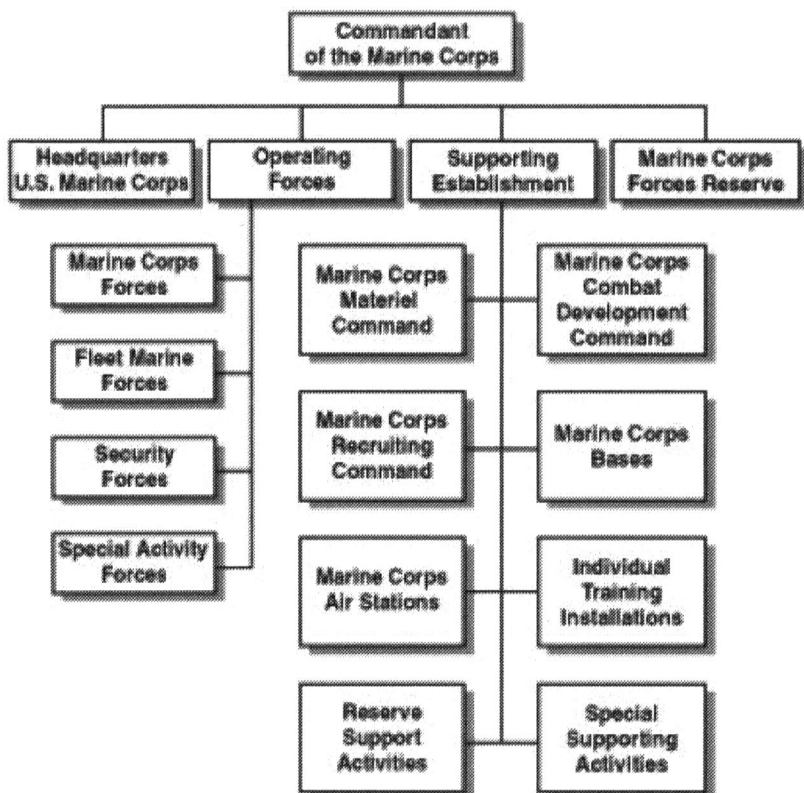

Figure 1-1. Marine Corps Organization.

- Preparing detailed instructions for the execution of approved plans and supervising the execution of those plans and instructions.
- Coordinating the actions of organizations of the Marine Corps.
- Performing such other duties, not otherwise assigned by law, as may be prescribed by the Secretary of the Navy or the Commandant.

Operating Forces

Assigned Marine Corps Forces

All Marine Corps combat, combat support, and combat service support units are part of the assigned Marine Corps forces. Normally, these forces are task-organized for employment as MAGTFs.

The Secretary of Defense "Forces for Unified Commands" memorandum assigns designated Marine Corps operating forces to Commander in Chief, U.S. Joint Forces Command (USCINCJFCOM) and Commander in Chief, U.S. Pacific Command (USCINCPAC). USCINCJFCOM exercises combatant command (command authority) or COCOM of II Marine Expeditionary Force (MEF) through the Commander, Marine Corps Forces, Atlantic (COMMARFORLANT). COMMARFORLANT has Service component responsibilities to USCINCJFCOM, U.S. Commander in Chief, Europe (USCINCEUR), and Commander in Chief, U.S. Southern Command (USCINCSO) as COMMARFORLANT, Commander, Marine Corps Forces, Europe (COMMARFOREUR), and Commander, Marine Corps Forces, South (COMMARFORSOUTH) respectively. USCINCPAC exercises COCOM of I and III MEF through the Commander, Marine Corps Forces, Pacific (COMMARFORPAC). COMMARFORPAC is also the Service component commander for Commander in Chief, U.S. Central Command (USCINCCENT) as Commander, Marine Corps Forces, Central (COMMARCENT). In addition to his Service component responsibilities for USCINCPAC and USCINCCENT, COMMARFORPAC has multiple responsibilities in Korea. He exercises Service component responsibilities over Marine Corps forces as Commander, U.S. Marine Corps Forces - Korea (COMUSMARFOR-K). He also exercises functional component responsibilities as Commander, Combined Marine Forces Command (COMCMFC).These assignments reflect the peacetime disposition of Marine Corps forces. MEFs are apportioned to the geographic combatant commanders for contingency planning and are provided to these combatant commands when directed by the Secretary of Defense.

Assigned Marine Corps forces are commanded by a combatant command-level Marine Corps component commander. He is responsible for—

- Training and preparing Marine Corps forces for operational commitment commensurate with the strategic situation and the combatant commander's requirements.
- Advising the combatant commander on the proper employment of Marine Corps forces, participating in associated planning, and accomplishing such operational missions as may be assigned.
- Providing Service administration, discipline, intelligence, and operational support for assigned forces.
- Identifying requirements for support from the Marine Corps supporting establishment.
- Performing such other duties as may be directed.

Fleet Marine Forces

Fleet Marine Forces units serve with Navy fleets in the seizure or defense of advanced naval bases and in the conduct of such land operations as may be essential to the prosecution of naval operations in support of the joint campaign. When assigned, Fleet Marine Force units are commanded by the Commanding Generals, Fleet Marine Force, Atlantic, Europe, South, or Pacific.

When the combatant commander tasks the Marine Corps component commander to provide assigned Marine Corps forces to the Navy component commander, the combatant command-level Marine Corps component commander CHOPs (change of operational control) MAGTFs and designated forces from Marine Corps forces to the Navy component commander. These Fleet Marine Forces then serve with a numbered fleet or for naval operations and other commitments; e.g., deployed Marine expeditionary units (MEUs).

The relationship between Marine Corps forces and the Fleet Marine Force reflects the roles and functions of the Marine Corps. The Marine Corps has separate responsibilities to provide forces for use by the combatant commanders—*Marine Corps forces*—and by Navy operational commanders—*Fleet Marine Forces*. See figure 1-2.

Security Forces

The 4th Marine Expeditionary Brigade (MEB) (Antiterrorism (AT)) provides the unified combatant commanders with a rapidly deployable and sustainable specialized antiterrorism force to *deter, detect, and defend against terrorist actions and conduct initial incident response* to combat the threat of terrorism worldwide. The 4th MEB (AT) provides the following capabilities:

- Chemical, biological, radiological, nuclear, and high explosive incident response.
- Physical and electronic security.
- Integrated vulnerability assessment and threat analysis.
- Explosive ordnance detection and disposal.
- Lethal and nonlethal weapons employment and training.
- Urban search and rescue.
- Physical security and antiterrorism/force protection training.

The 4th MEB (AT) deploys a forward command element (CE)/assessment team within 6 hours of notification and maintains a task-organized antiterrorism/incident response MAGTF on 12-hour alert. The entire MEB (AT) can deploy within

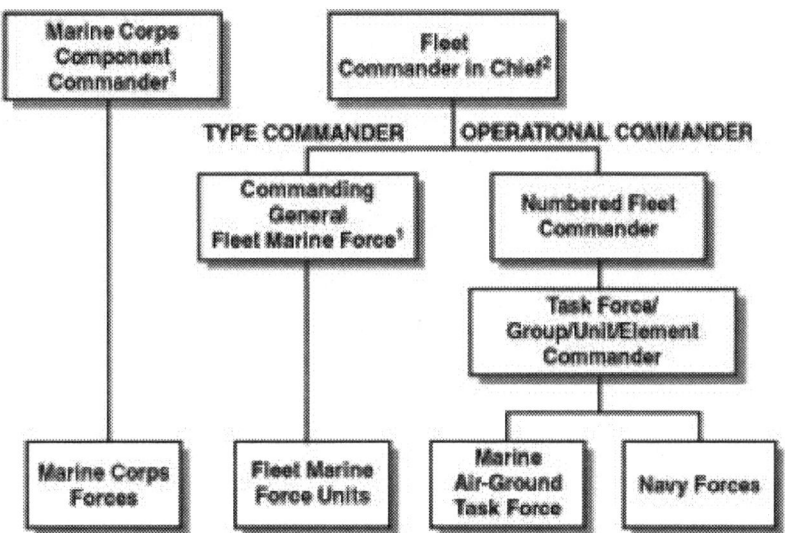

Figure 1-2. Marine Corps Forces and Fleet Marine Force Relationship.

72 hours of notification. It may include an air contingency battalion/antiterrorism battalion, a Chemical/Biological Incident Response Force, Marine Corps Security Force Battalion elements, and Marine Security Guard Battalion elements.

The Marine Corps Security Force Battalion provides armed antiterrorism and physical security trained forces to designated naval installations, vessels or units. The Battalion's Fleet Antiterrorism Security Team (FAST) companies provide Fleet Commanders in Chief (CINCs) and Fleet Commanders forward-deployed FAST platoons for responsive short-term security augmentation of installations, ships or vital naval and national assets when force protection conditions have been elevated beyond the capabilities of the permanent security forces. Marine Corps Security Force Battalion companies operate under operational control (OPCON) of the designated Navy commanding officer and under the administrative control (ADCON) of the Commandant of the Marine Corps through the Commanding General, Fleet Marine Force, Atlantic.

Special Activity Forces

Special activity forces provide security or services or perform other certain special type duties for agencies other than the Department of the Navy. Assignment of personnel to and the mission of these forces are specified by the supported agency and approved by the Commandant. The Marine Corps provides Marines from the Marine Security Guard Battalion to meet the security guard detachment requirements at foreign service posts throughout the world. The Marine security guard detachment mission is to provide internal security services to selected Department of State embassies, consulates, and legations to prevent the compromise of classified material and equipment and protect United States citizens and government property. Marine security guard detachments operate under the OPCON of the Secretary of State and under the ADCON of the Commandant of the Marine Corps via the Commander, Marine Corps Forces, Atlantic/Commanding General 4th MEB (AT).

Supporting Establishment

The supporting establishment assists in the training, sustainment, equipping, and embarkation of deploying forces. The supporting establishment includes—

- Marine Corps Materiel Command.
- Marine Corps Combat Development Command.
- Marine Corps Recruiting Command.
- Marine Corps bases.
- Marine Corps air stations.
- Individual training installations.
- Reserve support activities.

Marine Corps Forces Reserve

Marine Corps Forces Reserve is an integral part of *Marine Corps Total Force*. It is organized, trained, and equipped under the direction of the Commandant and commanded by the Commander, Marine Corps Forces Reserve. The Commander, Marine Corps Forces Reserve provides trained and qualified units and individuals to be available for active duty in time of war, national emergency, and at such other times as the national security may require. In recent years, the Marine Corps Forces Reserve has been increasingly called upon to provide peacetime operational support. This operational support enhances the entire Marine Corps' operational readiness and reduces the strain of the operational tempo on the active forces. Marine Corps Forces Reserve also maintains close contact with the American people through community outreach and operates reserve training centers that ensure forces are ready in the event of mobilization.

Like the active forces, it is a combined arms force with balanced ground, aviation, and combat service support units. The Marine Corps Forces Reserve includes a division, wing, and force service support group and unique capabilities such as civil affairs groups, aviation aggressor squadrons, and air-naval gunfire liaison companies. Reserve units routinely exercise with the active forces and are assigned operational responsibilities. Marine Corps Forces Reserve units and individuals are available for employment on short notice after mobilization and any required refresher training. They can provide augmentation, reinforcement or reconstitution of regular Marine Corps forces to satisfy mission requirements. The Marine Corps Forces Reserve shares the same commitment to expeditionary readiness as the active duty Marine Corps.

MARINE CORPS ETHOS

The Marine Corps' most important responsibility is to win the Nation's battles. Winning these battles rests, as it has throughout the Nation's history, on the individual Marine. The Marine Corps must recruit America's finest young men and women and make Marines capable of winning the Nation's battles and becoming outstanding citizens. Accordingly, the institutional focus of the Marine Corps is on how Marines are inculcated with the ethos of the Marine Corps, trained for combat, and equipped with the best possible weapons to win these battles. Marine Corps ethos are based on the core values of *honor*, *courage*, and *commitment*. These values provide a framework for how Marines act and think. Strict adherence to the core values, coupled with rigorous training and education, ensure a Marine Corps that is made up of men and women with intellectual agility, initiative, moral courage, strength of character, and a bias for action.

At the heart of the ethos of the Marine Corps is the belief that every Marine is first a rifleman; specialty skills such as being an aviator, a tanker, a radio operator or a mechanic are secondary. Marines understand that everything they do is ultimately to defeat the enemy. Marines fighting at the forward edge of the battle area (FEBA) are supported by Marines who distribute supplies, repair battle-damaged equipment, and evacuate wounded Marines. There are no "rear area" Marines, as all Marines are expected to drop their wrenches, leave their computers, and pick up their rifles to defend their position or form provisional units to reinforce an attack.

The belief that every Marine is a rifleman is instilled in Marines during their initial training and is continually reinforced through training exercises and formal schooling. All Marine officers attend The Basic School before receiving any military occupational specialty training. Staff noncommissioned officers and noncommissioned officers attend academies designed to promote Marine leadership skills regardless of military occupational specialty. This collective training is critical in ensuring all Marines embrace the Marine Corps ethos.

Marines possess a number of characteristics that contribute to success in all operations. These characteristics, based on the Marine Corps core values include—

- Offensive spirit.
- Loyalty.
- Discipline.
- Mental and physical fitness.
- Tactical and technical proficiency.
- Readiness.
- Expeditionary mindset.
- Selflessness.
- Esprit de corps.

Such characteristics are essential if Marines are to employ the Marine Corps' maneuver warfare philosophy on a chaotic battlefield. Emerging operational concepts like expeditionary maneuver warfare, with its supporting concepts, and the demands of MOOTW can best be accomplished by forces that are completely imbued with the ethos and traditions of the Marine Corps.

By making Marines who embrace this Marine Corps culture and exhibit an expeditionary mindset, building on a history of innovation and experimentation to develop new warfighting concepts, and capitalizing on emerging technologies, the Marine Corps continues to be the Nation's expeditionary force in readiness.

CHAPTER 2

Marine Corps Expeditionary Operations

Contents

Marine Corps Core Competencies	2-2
Warfighting Culture and Dynamic Decision-making	2-2
Expeditionary Forward Operations	2-2
Sustainable and Interoperable Littoral Power Projection	2-3
Combined Arms Integration	2-3
Forcible Entry from the Sea	2-3
Expeditionary Operations	2-4
Force Projection	2-5
Amphibious Operations	2-6
Amphibious Operation Command Relationships	2-6
Forcible Entry through an Amphibious Assault	2-8
Sequence	2-9
Maritime Pre-positioning Force	2-11
Sustained Operations Ashore	2-12
Enabling Force	2-12
Decisive Force	2-12
Exploitation Force	2-13
Sustaining Force	2-13
Emerging Concepts and Technologies	2-13
Expeditionary Maneuver Warfare	2-14
Operational Maneuver from the Sea	2-15
Ship-to-Objective Maneuver	2-16
Maritime Pre-positioning Force Future	2-17
Expeditionary Bases and Sites	2-18

> *"If a service does not possess a well-defined strategic concept, the public and the political leaders will be confused as to the role of the service, uncertain as to the necessity of its existence, and apathetic or hostile to the claims made by the service upon the resources of society."*
> —Samuel P. Huntington

> *". . . given the global nature of our interests and obligations, the United States must maintain its overseas presence forces and the ability to rapidly project power worldwide in order to achieve full spectrum dominance."*
> —Joint Vision 2020

MARINE CORPS CORE COMPETENCIES

The Marine Corps' contribution to national security and its role within a naval expeditionary force rest upon five unique core competencies. These competencies define the essence of the Marine Corps' institutional culture and its contribution to the national military establishment. Core competencies are the set of specific capabilities or activities fundamental to a Service or agency role. The Marine Corps' core competencies allow Marines to conduct expeditionary operations across the spectrum of crisis and conflict around the world. These core competencies, articulated by the Commandant of the Marine Corps in *Expeditionary Maneuver Warfare*, follow.

Warfighting Culture and Dynamic Decision-making

War is fundamentally a clash of human wills and, as such, its outcomes are often determined more by human qualities than by technology. For this reason, Marines focus on the force of human resolve and utilize technology to leverage the chaos and complexity of the battlefield. From early on, Marines are instilled with a determination to accomplish the mission, ingeniously adapting available resources in chaotic and austere operating environments to realize success in battle. While Marines leverage technology to enhance tempo and decision-making capabilities—their training, education, and experience foster decisiveness even in the absence of perfect information. This 'decision superiority' recognizes that technology will never fully obviate fog and friction, and that the human ability to make effective decisions in battle is best achieved by intuition built through rigorous training, practiced discipline, and relevant experience. Leveraging chaos and complexity through information and decision-making technologies provides a truly asymmetric advantage to Marines engaged across the spectrum of operations.

Expeditionary Forward Operations

Marines are continuously deployed around the world near potential trouble spots where they can deter aggression, respond quickly, and resolve crises whenever called. The Corps' naval character and its strategically mobile presence enhance cultural and situational awareness of potential operating areas. This enhanced awareness enables Marines to work with friends and allies throughout each region, and is a cornerstone of the CINCs' engagement plans.

Sustainable and Interoperable Littoral Power Projection

Today's scalable MAGTFs can access the world's littoral regions on short notice, responding quickly with a force tailored to the mission at hand. Our partnership with the Navy provides significant organic sustainment capabilities from the sea, and reduces the CINC's requirement to dedicate precious lift assets to sustaining early entry forces. This means that sustainable and credible naval forces can begin responding to a crisis early, supporting other elements of national power while giving the joint or combined commander time to develop the theater of operations more fully as required.

Combined Arms Integration

Marines pioneered development of concepts such as close air support and vertical envelopment. MAGTFs constantly blend the art and science of commanding, controlling, training, and executing combined arms operations from air, land, sea, and space. Marines understand the logic and synergy of joint and multinational forces under the 'Single Battle' concept because of their culture and training in combined arms and expeditionary operations. They have experience with other government and nongovernment agencies. MAGTFs deliver desired 'effects' through both lethal and nonlethal means, with simultaneity and depth across the spectrum of operations. Marine employment of combined arms at the tactical level of war is a truly unique capability that reflects our innovative approach to warfighting and complements the tenets of maneuver warfare.

Forcible Entry from the Sea

Together, the Navy and Marine Corps provide the Nation with its primary capability to rapidly project and sustain combat power ashore in the face of armed opposition. When access to safeguarding America's interests is denied or in jeopardy, forward-present, rapidly deployable Marine forces are trained and ready to create and exploit seams in an enemy's defenses by leveraging available joint and naval capabilities, projecting sustainable power ashore, and securing entry for follow-on forces. MEFs, reinforced by maritime pre-positioned assets when required, allow the United States to protect its worldwide interests, reassure allies, and fortify other elements of national power.

EXPEDITIONARY OPERATIONS

An expedition is a military operation conducted by an armed force to accomplish a specific objective in a foreign country. Expeditionary operations encompass the entire range of military operations, from humanitarian assistance to forcible entry in a major theater war. The defining characteristic of expeditionary operations is the projection of force into a foreign setting. Successful expeditionary operations require—

- **Expeditionary Mindset.** Expeditionary Marine forces must establish and maintain an expeditionary mindset—an expeditionary culture—devoted to readiness and the mental agility and adaptability to accommodate changing conditions and accomplish rapidly changing missions with the forces and capabilities at hand.
- **Tailored Forces.** Marine forces are task-organized into MAGTFs to conduct expeditionary operations. MAGTFs are designed to accomplish the mission assigned and do not include forces or capabilities not required by the mission. Therefore, those forces needed to do the job—and only those forces—are employed.
- **Forward Deployment.** The presence of forward-deployed MAGTFs close to the crisis or objective area can expedite accomplishing the mission. They allow for a real deterrence, as the threat of employment is imminent and credible. Forward-deployed MAGTFs also can serve as a precursor to larger follow-on forces.
- **Rapid Deployment.** Expeditionary forces must be able to get to the crisis or AO quickly with all their capabilities ready to be employed. MAGTFs can rapidly deploy using airlift, sealift or movement or maneuver from a forward expeditionary site. Marines are always prepared to deploy anywhere in the world.
- **Expeditionary Basing.** Marines are prepared to take advantage of any opportunity to use expeditionary basing or sites to support rapid deployment and employment within the AO. Amphibious shipping, forward expeditionary sites, and intermediate staging bases are all methods the MAGTFs can employ to ensure the rapid buildup and effective employment of combat power.
- **Forcible Entry.** Expeditionary forces must be able to gain access to the AO despite the efforts of the enemy to prevent it. While Marines strive to avoid enemy strengths and take advantage of the enemy's weakness, MAGTFs must be prepared to defeat the enemy to allow follow-on operations. Marines are highly trained in forcible entry techniques such as amphibious assaults and helicopterborne (air assault) operations. Marines also train with allied, multinational, and joint forces such as United Kingdom Royal Marines, Republic of Korea Marines, and United States Army airborne forces.

- **Sustainment.** Expeditionary operations are often conducted in austere theaters or undeveloped areas of the world. Forces must be able to sustain their operations, providing the essential supplies and services necessary to keep the force manned and equipped to accomplish the mission. MAGTFs are well-suited to operate in these conditions as MAGTFs bring robust logistic and combat service support to the operation. Sea-basing, expeditionary sites, and the use of pre-positioned supplies and equipment assist in sustaining the force.

FORCE PROJECTION

Forward-deployed MAGTFs, with their range of capabilities, are designed to enable the joint force commander to resolve crises and win conflicts. MAGTFs are uniquely suited to support the national security strategy by rapidly projecting the required capability into a foreign setting to abate the crisis. This capability is central to the United States ability to safeguard its national interests. Forward-deployed MAGTFs are prepared to meet a wide array of challenges in their AO. Their presence and engagement activities help to shape the crisis area. Finally, MAGTFs respond with appropriate force or capabilities to defeat the enemy, restore order or provide humanitarian relief.

The role of forward-deployed naval forces will grow even more critical as crises become more frequent, more unpredictable, and more difficult to resolve. Marines will be called upon to support, conduct, and in some cases, lead interagency crisis response operations. In such contingencies, the MAGTF's ability to establish *immediate* presence and access into the crisis area will be key. As the "first to respond" force, the MAGTF must be in or close to the intended AO, ready to gain access through forcible entry, and organized, trained, and equipped to respond to a wide variety of challenges. Deploying forces may be supported by a wide range of informational and security activities designed to demonstrate the Nation's commitment to regional security, strengthen ties with allies, and deter aggression. While these information operations may involve many agencies outside the MAGTF, it is the presence, readiness, and inherent flexibility of the MAGTF that ultimately projects national military power. In essence, this will enable the Marine Corps to expand its traditional "first to fight" role, becoming a "first to respond" element of national power.

The Marine Corps conducts force projection primarily through the use of MAGTFs conducting expeditionary operations employing three primary methods:

- Amphibious operations.
- MPF operations.
- Combination of the above methods.

Although the Marine Corps can deploy forces solely through organic, strategic, or theater air assets, our naval character favors the three primary methods.

AMPHIBIOUS OPERATIONS

Joint Pub (JP) 3-02, *Joint Doctrine for Amphibious Operations*, defines an amphibious operation as a military operation launched from the sea by an amphibious force embarked in ships or craft with the primary purpose of introducing a landing force ashore to accomplish the assigned mission. Amphibious operations require a high degree of training and specialized equipment to succeed. Marine Corps forces are specifically organized, trained, and equipped to deploy aboard, operate from, and sustain themselves from amphibious ships. They are specifically designed to project land combat power ashore from the sea.

Amphibious operations are normally part of a larger joint operation. JP 3-02 sets forth the doctrine and selected tactics, techniques, and procedures to govern amphibious operations. Under Title 10, United States Code, the Marine Corps has primary responsibility for developing LF doctrine, tactics, techniques, procedures, and equipment that are of common interest to the Army and the Marine Corps. The doctrine, tactics, techniques, and procedures for conducting amphibious operations are not immutable; they are dynamic and evolutionary, adapting to new technology, threats, and requirements. Command relationships, the use of specific operational areas, and the control of airspace and fires must be flexible and adaptable to make amphibious forces more responsive to the current environment and an uncertain future.

Amphibious Operation Command Relationships

Command relationships in amphibious operations should facilitate cooperative planning between the joint force, the Navy, and the Marine forces. They should ensure that appropriate responsibility and authority for the conduct of the amphibious operation is assigned to the commander of the Marine forces. The command relationship options available to a joint force commander or other establishing authority of an amphibious operation include OPCON, tactical control (TACON) and support as described in JP 0-2, *Unified Action Armed Forces (UNAAF)*, and JP 3-02. While doctrine should not specify a normal command relationship, typically a support relationship is established between the amphibious task force commander (Navy) and the LF commander (Marine or Army) based on the complementary capabilities of the amphibious task force and the LF. The establishing authority should consider the following factors when developing the establishing directive and designating the supported commander at various phases and events during the amphibious operation (not all inclusive):

- Responsibility for the preponderance of the mission.
- Force capabilities.
- Threat.
- Type, phase, and duration of the operation.
- Command and control capabilities.
- Battlespace assigned.
- Recommendations from subordinate commanders.

An amphibious force is an amphibious task force and a landing force together with other forces that are trained, organized, and equipped for amphibious operations. An amphibious task force (ATF) is a Navy task organization formed to conduct amphibious operations. The ATF, together with the landing force and other forces, constitutes the amphibious force. A landing force (LF) is a Marine Corps or Army task organization formed to conduct amphibious operations. The LF, together with the ATF and other forces, constitute the amphibious force. There are five types of amphibious operations, each designed to have a specific impact on the adversary.

Amphibious Assault

The principal type of amphibious operation. It involves establishing a force on a hostile or potentially hostile shore. The assaults on strongly defended islands of Tarawa and Iwo Jima during World War II are examples.

Amphibious Raid

An amphibious operation involving swift incursion into or the temporary occupation of an objective followed by a planned withdrawal. The raid on Makin Island by Marine Raiders against the Japanese in 1942 is an example.

Amphibious Demonstration

An amphibious operation conducted for the purpose of deceiving the enemy through a show of force intended to delude the enemy into adopting a course of action unfavorable to him. The 2d Marine Division conducted a highly successful demonstration off the island of Tinian in 1944, fixing the Japanese defenders in place while the actual landing occurred at an unexpected landing site elsewhere on the island.

Amphibious Withdrawal

An amphibious operation involving the extraction of forces by sea in naval ships or craft from a hostile or potentially hostile shore. The withdrawal of the 1st Marine Division's Marines, soldiers, and equipment along with 91,000 Korean refugees from the port of Hungnam following the retreat from the Chosin Reservoir in 1950 is a classic example of an amphibious withdrawal.

Other Amphibious Operations

The capabilities of amphibious forces may be especially suited to conduct other types of operations, such as noncombatant evacuation operations and foreign humanitarian assistance.

Forcible Entry through an Amphibious Assault

To support national interests and achieve national objectives, the United States must maintain the capability to gain access to a crisis area by forcible entry. United States forces cannot depend solely on the cooperation from contiguous countries to allow the introduction of forces within the crisis area. Forcible entry through an amphibious assault remains the Marine Corps' specialty. The amphibious assault is often considered the most difficult type of military operation due to the necessity to rapidly buildup combat power ashore. The unique power projection capabilities of the Marine Corps take advantage of the mobility provided by naval strategic lift and the organic logistics and sustainment within amphibious forces to provide significant capabilities to the joint force commander. Marine Corps amphibious assault forces complement other forcible entry forces such as airborne and air assault forces by seizing ports and airfields necessary to facilitate the arrival and sustainment of these forces.

Within the five types of amphibious operations, there are a number of tasks that the Marine Corps, as part of an amphibious force, can accomplish to facilitate naval and joint operations. The following are representative, but not inclusive, of tasks that may be performed:

- Attack an enemy operational or tactical center of gravity or critical vulnerability.
- Seize a lodgment, to include ports and airfields, for the introduction of follow-on forces.
- Seize areas for the development of advanced bases.
- Destroy, neutralize or seize enemy advanced bases and support facilities.
- Seize or conduct a preemptive occupation of areas that block free passage by opposing forces.

- Provide afloat strategic reserve to exploit opportunities and counter threats.
- Provide strategic deception to force the enemy to defend along littoral areas.
- Evacuate United States citizens and selected citizens from the host nation or third country nationals whose lives are in danger from a foreign country to a designated safe haven.
- Operate with and provide training to allied or coalition amphibious forces.
- Provide a secure environment until other forces arrive on-scene to allow humanitarian relief efforts to progress and facilitate the movement of food and medical care to relieve suffering and prevent the loss of life.

The adaptability and versatility of Marine Corps forces provide unique warfighting capabilities to the joint force commander. Amphibious forces are well-suited to accomplish a wide variety of operational missions. Opportunities should be sought to—

- Use the MAGTF decisively to seize or obtain the joint force commander's operational objectives.
- Use the full operational reach of the amphibious force, through its organic aviation, fires, and surveillance and reconnaissance assets.
- Use the command and control capability on amphibious ships to minimize staffs ashore and facilitate force protection.
- Sea-base logistics and selectively offload to minimize footprint ashore and facilitate force protection.

Once an amphibious force has executed its mission ashore, the LF can remain ashore to support ongoing land operations or may re-embark in amphibious shipping to be available for a new mission. Marine Corps forces can provide great flexibility to a joint force commander when they are re-embarked aboard amphibious shipping and poised to strike the enemy again.

Sequence

Amphibious operations generally follow distinct phases, though the sequence may vary.

When amphibious forces are forward-deployed or subsequent tasks are assigned, the sequence of phases may differ. Generally, forward-deployed amphibious forces use the sequence embarkation, planning, rehearsal (to include potential reconfiguration of embarked forces), movement to the operational area, and action. However, significant planning is conducted prior to embarkation to anticipate the most likely missions and to load assigned shipping accordingly.

The five phases of an amphibious operation are always required, but their sequence may change as circumstances dictate.

Planning

The planning phase normally denotes the period extending from the issuance of an order that directs the operation to take place and ends with the embarkation of landing forces. Planning, however, is continuous throughout the operation. Although planning does not cease with the termination of this phase, it is useful to distinguish between the planning phase and subsequent phases because of the change that may occur in the relationship between amphibious force commanders when the planning phase ends and the operational phase begins.

Embarkation

The embarkation phase is when the landing forces, with their equipment and supplies, embark in assigned shipping. Organization for embarkation needs to provide for flexibility to support changes to the original plan. The landing plan and scheme of maneuver ashore are based on conditions and enemy capabilities existing in the operational area before embarkation of the landing force. A change in conditions of friendly or enemy forces during the movement phase may cause changes in either plan with no opportunity for reconfiguration of the landing force. The extent to which changes in the landing plan can be accomplished may depend on the ability to reconfigure embarked forces.

Rehearsal

Rehearsal may consist of an actual landing or may be conducted as a command post exercise. The rehearsal phase is when the prospective operation is rehearsed to—

- Test the adequacy of plans, timing of detailed operations, and combat readiness of participating forces.
- Ensure that all echelons are familiar with plans.
- Provide an opportunity to reconfigure embarked forces and equipment.
- Verify communications for commonality, redundancy, security, and reliability.

Movement

The movement phase is when various elements of the amphibious force move from points of embarkation or from a forward-deployed position to the operational area. This move may be via rehearsal, staging or rendezvous areas. The movement phase is completed when the various elements of the amphibious force arrive at their assigned positions in the operational area.

Action

The decisive action phase is the period from the arrival of the amphibious force in the operational area through the accomplishment of the mission to the termination of the amphibious operation. While planning occurs throughout the entire operation, it is normally dominant prior to embarkation. Successive phases bear the title of the dominant activity taking place within the phase.

MARITIME PRE-POSITIONING FORCE

The MPF is an integral part of the Marine Corps' expeditionary capability. Rapid response to regional contingencies is its primary role. An MPF consists of the MPS squadron (MPSRON), Navy support element, and MAGTF fly-in echelon. Together they provide the joint force commander with a proven, flexible force that can quickly respond to a full range of missions from combat to humanitarian relief. Fundamental to the MPF is its interoperability with joint forces and its rapid introduction of combat forces into austere environments.

Comprised of specially designed ships, organized into three squadrons, MPSRONs carry equipment and supplies for 30 days of combat operations by a Marine expeditionary brigade (MEB) of approximately 16,000 Marines and Sailors. When deployed together, these squadrons provide equipment and supplies to support a MEF. These squadrons are forward-deployed to ensure rapid closure to the crisis area within a 5–14 day sailing period. See figure 2-1. MAGTF and Navy support element personnel are airlifted to a previously seized lodgment, a benign or host nation port and airfield or to an intermediate support

Figure 2-1. Current Location of Maritime Pre-positioning Ships Squadrons.

base where they link up with equipment and supplies offloaded from the MPSRON. If a port is not available, the squadron may be offloaded in-stream. A unique characteristic of the MPF is that the embarked equipment is maintained aboard ship and is combat-ready immediately upon offload. The entire squadron or selected capability sets from designated ships can be offloaded to support a wide range of MAGTF missions.

Movement of Marine forces and their combat essential equipment must be fully integrated with ongoing tactical operations ashore. The MPF, consisting of the MPS and the fly-in echelon of Navy and Marine forces, can then integrate with the MAGTF commander's scheme of maneuver while delivering combat service support and force sustainment, enabling a rapid force buildup or providing support and sustainment for a prolonged period from offshore. MPFs can tailor support packages to accommodate a variety of missions to include MOOTW of varying scope and complexity.

SUSTAINED OPERATIONS ASHORE

The Marine Corps also has the capability to operate independent of the sea to support sustained land operations ashore with the Army or coalition partners. The Marine Corps conducts sustained operations ashore to provide the joint force commander four options when fighting a land operation.

Enabling Force

The enabling force sets the stage for follow-on operations by other joint force components. The amphibious landing and subsequent operations ashore against the Japanese on Guadalcanal in 1942 set the stage for the arrival of Army forces to complete the seizure of the island in 1943. These enabling actions are not limited to the opening phases of the campaign such as establishing a lodgment, but may be conducted to divert attention away from the main effort. An example of this would be the role of I MEF in Operation Desert Storm (1991) in fixing the Iraqi forces in Kuwait while allowing Central Command's main effort, U.S. Army VII Corps, to maneuver to envelop the enemy.

Decisive Force

The decisive force exploits its advanced command and control system to identify gaps necessary to conduct decisive operations and reduce enemy centers of gravity. Decisive actions run the gamut from destruction of enemy military units to interdiction of critical lines of communications to the evacuation of American and third country nationals from untenable urban areas. An example of such a decisive action is the landing at Inchon in 1950 that severed the North Korean lines of communications and forced their withdrawal from South Korea.

Exploitation Force

The exploitation force takes advantage of opportunities created by the activity of other joint force components. The joint force commander may exploit these opportunities through rapid and focused sea-based operations by the MAGTF that capitalize on the results of ongoing engagements to achieve decisive results. The 24th MEU served in this role during operations to seize Grenada and safeguard American citizens in 1983. While Army forces fixed the Cuban and Grenadian forces at one end of the island, the Marines landed at will and maneuvered freely around the island, accomplishing the joint force commander's objectives.

Sustaining Force

The sustaining force maintains a presence ashore over an extended period of time to support continued operations by the joint force commander within the joint AO. Also includes providing logistical sustainment to joint and coalition forces until theater level sustainment is established. I MEF fulfilled this role in the early days of Operation Desert Shield (1990) in Saudi Arabia and Operation Restore Hope (1992–1993) in Somalia by providing sustainment to joint and Army forces until arrangements for theater support were complete.

EMERGING CONCEPTS AND TECHNOLOGIES

Concepts enable decisionmakers the ability to identify capabilities and changes to doctrine, organization, training, and education to create a force for the future. Future operational concepts are general descriptions of how military forces intend to fight in the future. Services and the joint community have relied increasingly on operational concepts as the "engines" for their combat development processes. These concepts also furnish the intellectual basis for experimentation and force development.

Increasing technological advancements will expedite the creation of capabilities articulated in emerging concepts. Technologically advanced systems will enhance the United States military forces' ability to exploit critical vulnerabilities and rapidly defeat centers of gravity. These centers of gravity will be located and identified using modern sensors and sophisticated intelligence collection and analysis. Some of the advanced systems that will provide the enhanced capability to conduct operations are advanced sensors and information systems; tilt rotor aircraft; vertical and short take-off aircraft; air-cushion vehicles and other hovercraft; high-speed shallow-draft ships; and the advanced amphibious assault vehicle.

EXPEDITIONARY MANEUVER WARFARE

Expeditionary maneuver warfare is the Marine Corps capstone concept. It prepares the Marine Corps as a "total force in readiness" to meet the challenges and opportunities of a rapidly changing world. Expeditionary maneuver warfare focuses our core competencies, evolving capabilities, and innovative concepts to ensure that the Marine Corps provides the joint force commander with forces optimized for forward presence, engagement, crisis response, and warfighting. Expeditionary maneuver warfare serves as the basis for influencing the Joint Concept Development and Experimentation Process and the Marine Corps Expeditionary Force Development System. It further refines the broad "axis of advance" identified in *Marine Corps Strategy 21* for future capability enhancements. In doing so, expeditionary maneuver warfare focuses on—

- Joint/multinational enabling. Marine forces are ready to serve as the lead elements of a joint force, act as joint enablers and/or serve as joint task force or functional component commanders (JFLCC, JFACC or JFMCC).
- Strategic agility (rapidly and fluidly transitioning from pre-crisis state to full operational capability in a distant theater [requires uniformly ready forces, sustainable and easily reorganized for multiple missions or functions]). They must be agile, lethal, swift in deployment, and always prepared to move directly to the scene of an emergency or conflict.
- Operational reach (projecting and sustaining relevant and effective power across the depth of the battlespace).
- Tactical flexibility (operating with tempo and speed and bringing multi-role flexibility [air, land, and sea] to the joint team).
- Support and sustainment (providing focused logistics to enable power projection independent of host-nation support and against distance objectives across the breadth and depth of a theater of operations).

These capabilities enhance the joint force's ability to reassure and encourage our friends and allies while we deter, mitigate or resolve crises through speed, stealth, and precision.

Expeditionary maneuver warfare focuses our warfighting concepts toward realizing the *Marine Corps Strategy 21* vision of future Marine forces with enhanced expeditionary power projection capabilities. It links Marine Corps concepts and vision for integration with emerging joint concepts. As our capstone concept, expeditionary maneuver warfare will guide the process of change to ensure that Marine forces remain *ready*, *relevant*, and *fully capable* of supporting future joint operations.

Operational Maneuver from the Sea

OMFTS applies across the range of military operations, from major theater war to smaller-scale contingencies. OMFTS applies maneuver warfare to expeditionary power projection in naval operations as part of a joint or multinational campaign. OMFTS allows the force to exploit the sea as maneuver space while applying combat power ashore to achieve the operational objectives. It reflects the Marine Corps' expeditionary maneuver warfare concept in the context of amphibious operations from a sea base, as it enables the force to—

- Shatter the enemy's cohesion.
- Pose menacing dilemmas.
- Apply disruptive firepower.
- Establish superior tempo.
- Focus efforts to maximize effect.
- Exploit opportunity.
- Strike unexpectedly.

The force focuses on an operational objective, using the sea as maneuver space to generate overwhelming tempo and momentum against enemy critical vulnerabilities. OMFTS provides increased operational flexibility through enhanced capabilities for sea-based logistics, fires, and command and control. Sea-basing facilitates maneuver warfare by eliminating the requirement for an operational pause as the LF builds combat power ashore and by freeing the MAGTF from the constraints of a traditional beachhead. OMFTS is based on six principles:

- **Focus on the Operational Objective.** The operation must be viewed as a continuous event from the port of embarkation to the operational objective ashore. Everything the force does must be focused on achieving the objective of the operation and accomplishing of the mission. Intermediate objectives or establishing lodgments ashore assume less importance in OMFTS as the force is centered on decisive maneuver to seize the force objective.
- **Use the Sea as Maneuver Space.** Naval forces use the sea to their advantage, using the sea as an avenue of approach and as a barrier to the threat's movement. This allows the force to strike unexpectedly anywhere in the littorals and to use deception to mislead the enemy as to actual point of attack.
- **Generate Overwhelming Tempo and Momentum.** The objective of maneuver warfare is to create a tempo greater than that of the enemy. The tempo generated through maneuver from the sea provides the commander freedom of action while limiting the enemy's freedom of action.

- **Pit Friendly Strength Against Enemy Weakness.** The commander identifies and attacks critical vulnerabilities where the enemy is weak, rather than attacking his center of gravity when it is strong.
- **Emphasize Intelligence, Deception, and Flexibility.** Deception enhances force protection while reconnaissance and intelligence are essential in identifying fleeting opportunities.
- **Integrate all Organic, Joint, and Multinational Assets.** To realize the maximum effectiveness the commander must ensure the coordinated use of all available forces and capabilities..

> OMFTS should not be viewed as a revolutionary new way of conducting amphibious and MPF operations. The brilliant amphibious operation at Inchon in 1950 is a classic example of OMFTS. It is an evolutionary way of using expeditionary forces assisted by greatly increased enhancements to current capabilities such as sea-basing. These enhancements will be in the form of new doctrine, organization, training and education, equipment, and technology.

When operating as part of a naval expeditionary force, MEFs will normally focus on conducting operations using OMFTS. The Marine commander, in concert with his Navy counterpart and higher-level direction, will orchestrate the employment of amphibious forces, MPFs, and Marine forces operating from land bases to shape events and create favorable conditions for future combat actions. The amphibious forces will normally execute tactical-level maneuver from the sea to achieve decisive action in battle. For the action to be decisive, the battle must lead to the achievement of the operational objectives.

Ship-to-Objective Maneuver

STOM is the tactical implementation of OMFTS by the MAGTF to achieve the joint force commander's operational objectives. It is the application of maneuver warfare to amphibious operations at the tactical level of war. STOM treats the sea as maneuver space, using the sea as both a protective barrier and an unrestricted avenue of approach. While the aim of ship-to-shore movement was to secure a beachhead, STOM thrusts Marine Corps forces ashore at multiple points to concentrate at the decisive place and time in sufficient strength to enable success. This creates multiple dilemmas too numerous for the enemy commander to respond to, disrupting his cohesiveness and diminishing his will or capacity to resist. This concept focuses the force on the operational objective, providing increased flexibility to strike the enemy's critical vulnerabilities. Sea-basing of some of the fire support and much of the logistics support reduces the footprint of forces ashore while maintaining the tempo of operations. Emerging command and control capabilities will allow commanders to control the maneuver of their

Marine Corps Operations

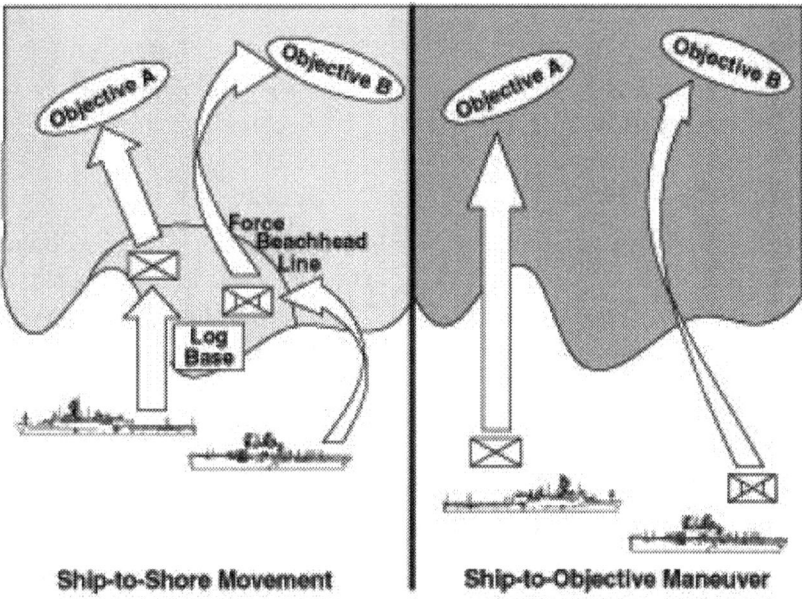

Figure 2-2. Operational Objectives.

units the moment they cross the line of departure at sea, to include changing the axis of advance or points where they cross the beach during the assault.

In STOM, rather than an amphibious assault to establish a force on a hostile or potentially hostile shore, an *amphibious attack* may occur. An amphibious attack may be defined as an attack launched from the sea by amphibious forces directly against an enemy operational or tactical center of gravity or critical vulnerability. See figure 2-2.

Maritime Pre-positioning Force Future

MPF Future is the concept that next-generation MPFs will contribute to forward presence and power projection capabilities, which will remain central to United States deterrence and conflict resolution strategies well into the future. Enhancements envisioned in MPF Future will expand the use of the future MPF across an increased range of contingencies. The concept is best illustrated through the pillars of future MPF operations:

- **Force Closure**—provides for at-sea arrival and assembly of forces.
- **Amphibious Force Integration**—using selective offload capabilities to reinforce the assault echelon of an amphibious force.

- **Indefinite Sustainment**—by serving as a sea-based conduit for logistic support.
- **Reconstitution and Redeployment**—without a requirement for extensive material maintenance or replenishment at a strategic sustainment base.

Expeditionary Bases and Sites

Future contingencies will compel an ever-increasing reliance on expeditionary bases and sites, especially sea-basing, to support and sustain expeditionary maneuver warfare. Expeditionary bases and sites are locations in and out of the AO that can support the deployment, employment, and sustainment of expeditionary forces. They might include—

- Intermediate staging bases outside of the AO.
- Sites located within friendly contiguous nations or the host nation.
- Expeditionary airfields and forward operating bases established within the AO by the expeditionary forces.
- Existing facilities within the AO seized from the enemy.
- United States military bases overseas or in CONUS located near the AO.
- Amphibious shipping.
- MPS squadrons.

Marine Corps forces can quickly establish these temporary and often austere expeditionary bases and sites providing the ability to project, support, and sustain forces. Amphibious shipping or the MPS allows Marine Corps forces to operate from a mobile sea base far from the enemy's shore. Sea-basing allows the Marine Corps to bring ashore only those forces and assets essential to the mission. This provides the joint force commander with increased operational freedom, precluding the need to establish, man, and protect extensive shore-based facilities. The ability to conduct logistics and sustainment activities from the sea base, existing infrastructure ashore, or any combination will reduce the footprint ashore, thereby minimizing the threat to deployed forces.

CHAPTER 3

Marine Corps Forces

Contents	
Marine Corps Component.	3-1
Role and Responsibilities to the Commandant	3-3
Role and Responsibilities to a Combatant Commander.	3-3
Role and Responsibilities to the Joint Force Commander	3-4
Joint Operations Conducted Through Service Component Commanders	3-5
Joint Operations Conducted Through Functional Component Commanders.	3-5
The Marine Corps Component Commander as a Functional Component Commander.	3-7
Role and Responsibilities to the MAGTF Commander.	3-9
Component Command Relationships and Staff Organization	3-10
The Marine Air-Ground Task Force .	3-10
Capabilities.	3-11
Elements	3-12
Supporting Establishment	3-15
Types .	3-16
Marine Logistics Command .	3-20

> *"I have just returned from visiting the Marines at the front, and there is not a finer fighting organization in the world."*
>
> —General Douglas MacArthur, USA

The Marine Corps organizes its operational forces as Marine Corps components and as MAGTFs to provide task-organized, self-sustaining, multipurpose forces to the joint force or naval expeditionary force. These uniquely organized Marine Corps forces can respond to a wide range of operational and tactical missions and tasks, providing the National Command Authorities with an unmatched combination of deployment and employment options.

MARINE CORPS COMPONENT

The Marine Corps will normally conduct operations as part of a joint force. While the overwhelming majority of operations that Marine Corps forces will be involved with will be joint, there may be instances where Marine Corps forces may conduct single-Service operations. A combatant commander can establish command structure and conduct operations using a single-Service force. See JP 0-2 and JP 3-0, *Doctrine for Joint Operations*, for more information. Joint forces are constituted with subordinate organizations known as components. Per JP 0-2, the Service forces that comprise the joint force operate as components. Normally a joint force is organized with a combination of Service and functional components. Regardless of how a joint force commander organizes his forces, if Marine Corps forces are assigned, there is a Marine Corps component. There are two levels of Marine Corps components: a Marine Corps component under a unified command and a Marine Corps component under a subordinate unified command or a joint task force. The Marine Corps component commander deals directly with the joint force commander in matters affecting Marine Corps forces. See figure 3-1.

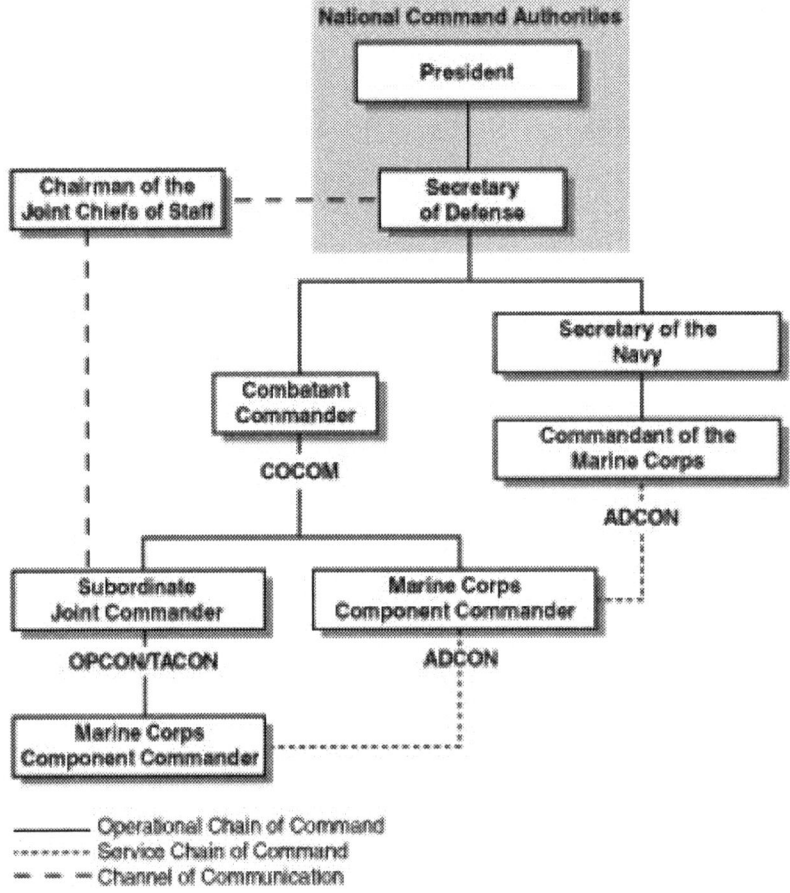

Figure 3-1. Chains of Command and Command Relationships.

The Marine Corps component commander commands, trains, equips, and sustains Marine Corps forces. He sets the conditions for their success in the battlespace. He translates the joint force commander's intent into Marine Corps forces' actions.

Role and Responsibilities to the Commandant

The Marine Corps component commander is responsible for and accountable to the Commandant for the internal discipline, training, and administration of his forces. His responsibilities specifically include—

- Internal discipline and administration.
- Training in Marine Corps doctrine, tactics, techniques, and procedures.
- Logistic functions normal to the command, except as otherwise directed by higher authority.
- Service intelligence matters and oversight of intelligence activities according to current laws, policies, and directives.

While the Marine Corps component commander responds to the joint force commander in the operational chain of command, his component is equipped, manned, and supported by the Commandant through the Service chain of command. The Commandant's relationship with the Marine Corps component commander is through the Service chain—not the operational chain. Unless otherwise directed by the combatant commander, the Marine Corps component commander will communicate through the combatant command on those matters that the combatant commander exercises COCOM or directive authority. On Service-specific matters—personnel, administration, and unit training—the Marine Corps component commander will normally communicate directly with the Commandant, informing the combatant commander as the combatant commander directs.

A combatant command-level Marine Corps component is generally required for a major theater war. A subordinate joint command-level Marine Corps component is normally appropriate for a smaller-scale contingency and MOOTW. MOOTW may occur simultaneously with a major theater war or a smaller-scale contingency.

Role and Responsibilities to a Combatant Commander

There are five combatant command-level Marine Corps components-Marine Corps Forces, Atlantic (MARFORLANT), Marine Corps Forces, Pacific (MARFORPAC), Marine Corps Forces, Europe (MARFOREUR),

MARFORSOUTH, and Marine Corps Forces, United States Central Command (MARFORCENT). Only MARFORLANT and MARFORPAC have assigned forces. COMMARFORLANT is the combatant command-level Marine Corps component commander for the USCINCJFCOM and is assigned as the combatant command-level Marine Corps component commander to both USCINCEUR, and USCINCSO. COMMARFORPAC is the combatant command-level Marine Corps component commander for USCINCPAC, and is designated as the combatant command-level Marine Corps component commander to the USCINCCENT. COMMARFORPAC is also designated as COMUSMARFOR-K to the Commander United States Forces Korea, a subordinate unified commander. See Marine Corps Doctrinal Publication (MCDP) 1-0.1, *Componency*, for more information.

Role and Responsibilities to the Joint Force Commander

The joint force commander conducts campaigns through a series of related operations. He conducts his campaigns by assigning component commanders missions that accomplish strategic and operational objectives. The orientation of the Marine Corps component commander is *normally* at the operational level of war, while the MAGTF commander is *normally* at the tactical level. See figure 3-2. Naturally, there is some overlap. The Marine Corps component commander is normally responsible to set the conditions for Marine Corps tactical operations. These operations include military actions executed by the MAGTF, other assigned or attached Marine Corps forces, and assigned or attached forces from other Services and nations.

A joint force commander organizes his forces to accomplish the assigned mission based on the factors of mission, enemy, terrain and weather, troops and support available, and time available (METT-T), and the concept of operations.

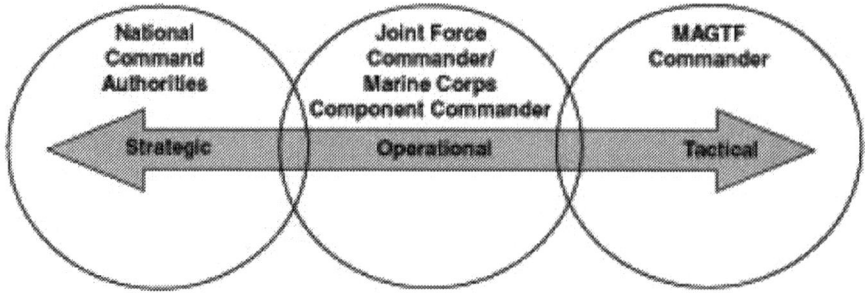

Figure 3-2. Commander's Level of War Orientation.

The organization should provide for unity of effort, centralized planning, and decentralized execution. The joint force commander establishes subordinate commands, assigns responsibilities, and establishes appropriate command and support relationships. He should allow Service tactical and operational assets to function generally as they were designed, trained, and equipped. The intent is to meet the needs of the joint force while maintaining the tactical and operational integrity of the Service organizations. He can organize and conduct operations through Service component commanders, functional component commanders or a combination of the two.

Joint Operations Conducted Through Service Component Commanders

A joint force commander may conduct operations through the Service component commanders. Conducting operations through Service components has certain advantages, including clear and uncomplicated command lines. This relationship is appropriate when stability, continuity, economy, ease of long-range planning, and scope of operations dictate preserving the organizational integrity of Service forces. These conditions apply when most of the required functions in a particular dimension are unique to a single-Service force or when Service force capabilities or responsibilities do not significantly overlap. In addition, Service component commands provide administrative and logistic support for their forces in a joint operation.

When the joint force commander conducts joint operations through Service component commanders, the Marine Corps component commander and the other Service component commanders have command—OPCON and ADCON—of their assigned Service forces. The joint force commander may also establish a support relationship between Service components to facilitate operations. Support is a command authority. A superior commander establishes a support relationship between subordinate commanders when one should aid, protect, complement or sustain the other. The four categories of support are general, mutual, direct, and close. See MCDP 1-0.1, *Componency*.

Joint Operations Conducted Through Functional Component Commanders

A joint force commander may conduct operations through functional components or employ them primarily to coordinate selected functions. Regardless of how the joint force commander organizes his assigned or attached forces, a Marine Corps component is included to provide administrative and logistic support for the assigned or attached Marine Corps forces. See figure 3-3 on page 3-6. Functional components may be established across the range of military operations to

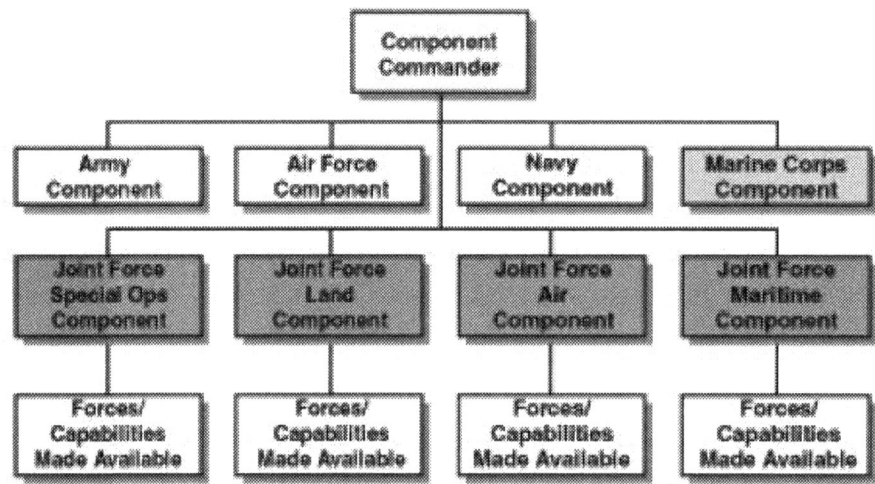

Figure 3-3. Combatant Command Organized by Functional Components.

perform operational missions that may be of short or extended duration. Functional components can be appropriate when forces from two or more Military Departments must operate in the same dimension or medium or there is a need to accomplish a distinct aspect of the assigned mission. *Functional components are components of a joint force and do not constitute a "joint force" with the authorities and responsibilities of a joint force.*

When the joint force commander centralizes direction and control of certain functions or types of joint operations under functional component commanders he must establish the command relationships. The joint force commander must designate the military capability that will be made available for tasking by the functional component commander and the appropriate command relationship(s) the functional component commander will exercise. For example, a joint force special operations component commander normally has OPCON of assigned forces and a joint force air component commander is normally delegated TACON of air defense, long-range interdiction, or long-range reconnaissance sorties or other military capability made available. The policy for the command and control of Marine Corps aviation, specifically covered by the Chairman of the Joint Chiefs of Staff "Policy for Command and Control of USMC Tactical Air in Sustained Operations Ashore," is found in JP 0-2.

The Marine Corps component commander retains command—OPCON and ADCON—of those Marine Corps forces and capabilities not designated by the joint force commander for tasking by functional component commanders. The

Marine Corps component commander advises functional component commanders on the most effective use of Marine Corps forces or capabilities made available. Marine Corps forces or capabilities made available by the joint force commander respond to the functional component commander for operational matters based on the existing command relationship. All Marine Corps forces receive administrative and logistic support from the Marine Corps component commander. The joint force commander may also establish a support relationship between components to facilitate operations. See MCDP 1-0.1 for more information on the designation and responsibilities of functional component commanders.

Designation of a functional component commander must not affect the command relationships between Service component commanders and the joint force commander. The joint force commander must specifically assign the responsibilities and authority of the functional component commander. He defines the responsibilities and authority based on the concept of operations and may alter these responsibilities and authority during the course of an operation. Functional component commander responsibilities are found in JP 0-2 and JP 3-0, *Doctrine for Joint Operations*.

The Marine Corps Component Commander as a Functional Component Commander

Forward-deployed naval forces, including Marine Corps forces, are usually the first conventional forces to arrive in an austere theater or AO during expeditionary operations. The Marine Corps component commander's inherent capability to command and control Marine Corps forces—and attached or assigned forces of other Services or nations—allows him to command and control a functional component. The Marine Corps component commander can serve as a functional component commander in most smaller-scale contingencies and MOOTW. If the Marine Corps component commander is assigned functional component commander responsibilities, execution is normally accomplished by the assigned MAGTF.

As the commander of the force most capable of rapid worldwide deployment, the Marine Corps component commander may serve as a functional component commander in the initial phase of a major theater war. As the theater matures and additional United States forces flow into the theater, the Marine Corps component commander's capability to command and control these joint forces diminishes. When the functional component commanders establish their headquarters and supporting infrastructure, they begin to assume command and control of their assigned forces and capabilities. The transition of functional component duties from the Marine Corps component commander continues until the gaining functional component commander demonstrates full operational capability. The joint force commander can designate the Marine Corps component commander as follows.

Joint Force Maritime Component Commander

This commander is responsible for planning, coordinating, and executing joint maritime operations. Joint maritime operations are performed with maritime capabilities or forces made available by components to support the joint force commander's operation or campaign objectives or to support other components of the joint force. The maritime environment includes oceans, littorals, riverine areas, and amphibious objective areas, and the coordinated airspace above them as defined by the joint force commander. The joint force commander will designate the component commander best suited to accomplish the mission as the joint force maritime component commander. When maritime operations are focused on littoral operations—and Marine Corps forces have the preponderance of the mission or capabilities to accomplish the mission—the Marine Corps component commander may be designated the joint force maritime component commander.

Joint Force Land Component Commander

This commander is responsible for planning, coordinating, and executing joint land operations. Joint land operations are performed with land capabilities or forces made available by components to support the joint force commander's operation or campaign objectives or to support other components of the joint force. Marine Corps component commanders normally have the preponderance of land forces and the necessary command and control capability to direct their activities during expeditionary operations in a smaller-scale contingency. In the early stages of a major theater war, the Marine Corps component commander may serve as the joint force land component commander, but as forces continue to build up in theater, the joint force commander will normally designate the Army Service component commander as the joint force land component commander.

Joint Force Air Component Commander

This commander is responsible for planning, coordinating, and executing joint air operations. Joint air operation are performed with air capabilities or forces made available by components to support the joint force commander's operation or campaign objectives or to support other components of the joint force. The expeditionary nature of Marine aviation and its associated command and control capability allow the Marine Corps component commander to function as the joint force air component commander in a smaller-scale contingency. In the early stages of a major theater war, the Marine Corps component commander may serve as the joint force air component commander, but as forces continue to buildup in theater, the joint force commander will normally designate another component commander as the joint force air component commander.

Most often, the joint force commander conducts operations through a combination of Service and functional component commands with operational responsibilities. Joint forces organized with Army, Navy, Air Force, and Marine Corps components will have special operations forces (if assigned) organized as a functional component. The joint force commander defines the authority and responsibilities of the Service and functional component commanders. However, the Service responsibilities, i.e., administrative and logistic, of the components must be given due consideration by the joint force commander.

In addition to functional component responsibilities, a joint force commander can assign the Marine Corps component commander other joint responsibilities. The joint force commander can designate the Marine Corps component commander as the area air defense commander, airspace control authority, joint rear area coordinator or to establish the joint search and rescue center.

Role and Responsibilities to the MAGTF Commander

The Marine Corps component commander is responsible for the employment of his forces and to support other component commanders as directed by the joint force commander. The Marine Corps component commander sets conditions for the successful employment of the MAGTF by ensuring that appropriate missions, forces, resources, battlespace, and command relationships are assigned or made available to the MAGTF. While principally a force provider and sustainer, the Marine Corps component commander may be assigned some operational responsibilities. He focuses on the formulation and execution of the joint force commander's plans, policies, and requirements. He coordinates strategic and operational actions with other component commanders to achieve unity of effort for the joint force. He accomplishes any assigned mission by executing Marine Corps component operations through the MAGTF and other assigned forces.

During employment, the command relationship between the Marine Corps component commander and the MAGTF commander can vary with each phase of an operation. The MAGTF commander may have command relationships with two types of components: functional and Service. When the MAGTF is OPCON or TACON to a functional component commander, the functional component commander provides the tasks and purpose for the MAGTF, which in turn drive the development of the MAGTF's course of action and subsequent planning efforts. If the joint force is organized on a Service component basis, the Marine Corps component commander provides the tasks and purpose for the MAGTF.

As the Service component commander, the Marine Corps component commander represents MAGTF interests at various joint force boards. He will participate on joint force boards along with any functional component commander the MAGTF may be supporting. Consequently, the MAGTF must keep the Marine Corps component commander informed of operational matters to ensure relevant and contextual representation at the various joint boards.

Component Command Relationships and Staff Organization

This difference in orientation is the result of the joint force commander's organization of forces and each subordinate commander's place in the operational chain of command and the assigned mission. This placement, in turn, determines the people and agencies with whom the Marine Corps component and MAGTF commanders must interact.

The Marine Corps component commander—who translates strategic objectives into operational objectives—must interact up the chain of command with the joint force commander, laterally with other component commanders, and down to his MAGTF commander. The MAGTF commander—who translates operational objectives into tactical actions—must interact up the chain of command with the Marine Corps component commander, laterally with adjacent tactical commanders, and down to his subordinate commanders. The Marine Corps component commander assigns the MAGTF commander missions that may accomplish objectives at both the operational and tactical levels of war when the joint force is organized on a Service component basis.

The difference in orientation of the Marine Corps component commander and the MAGTF commander also has an important influence on the Marine Corps component-MAGTF command relationship and the staff organization adopted by the Marine Corps component commander. The Marine Corps component-MAGTF command relationship and staff organization that the Marine Corps component commander selects depends on the mission, size, scope, and duration of the operation and the size of the assigned force. Three possible command relationships and staff organizations are: one commander with one staff, one commander with two staffs, and two commanders and two staffs. See MCDP 1-0.1.

THE MARINE AIR-GROUND TASK FORCE

The Marine Corps task-organizes for operations consistent with its statutory tasking to ". . . provide forces of combined arms, including aviation . . ." by forming MAGTFs. The MAGTF is a balanced, air-ground combined arms task organization of Marine Corps forces under a single commander, structured to accomplish a specific mission. It is the Marine Corps' principal organization for all missions across the range of military operations. It is designed to fight, while having the ability to prevent conflicts and control crises. All MAGTFs are task-organized and vary in size and capability according to the assigned mission, threat, and battlespace environment. See figure 3-4, page 3-12. They are specifically tailored for rapid deployment by air or sea and ideally suited for a forward presence role. A MAGTF provides the naval, joint or multinational commander with a readily available force capable of operating as—

- The landing force of an amphibious task organization.
- A land force in sustained operations ashore.
- A land force or the landward portion of a naval force conducting MOOTW such as noncombatant evacuations, humanitarian assistance, disaster relief or the tactical recovery of an aircraft or aircrew.
- A forward-deployed force providing a strong deterrence in a crisis area.
- A force conducting training with allied forces as part of a theater engagement plan.

Capabilities

MAGTFs provide joint force commanders with the capability to—

- Move forces into crisis areas without revealing their exact destinations or intentions.
- Provide continuous presence in international waters.
- Provide immediate national response in support of humanitarian and natural disaster relief operations.
- Provide credible combat power in a nonprovocative posture, just over the horizon of a potential adversary, for rapid employment as the initial response to a crisis.
- Support diplomatic processes for peaceful crisis resolution before employing immediately responsive combat forces.
- Project measured degrees of combat power ashore, day or night, and under adverse weather conditions, if required.
- Introduce additional forces sequentially into a theater of operations.
- Operate independent of established airfields, basing agreements, and overflight rights.
- Conduct operations ashore using organic combat service support brought into the AO.

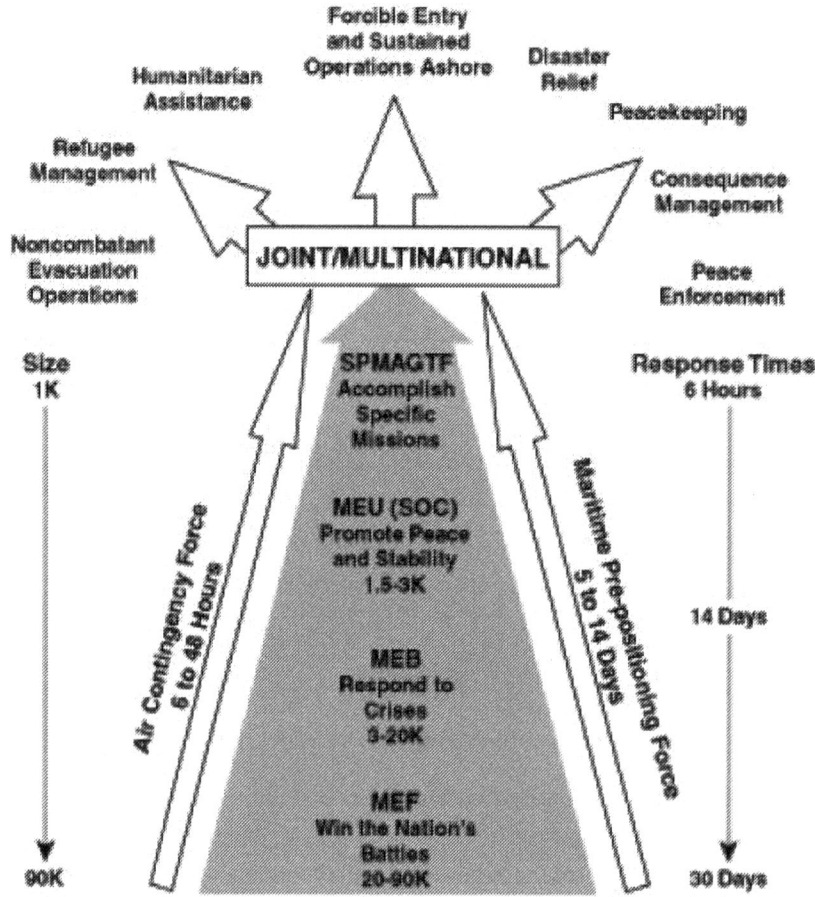

Figure 3-4. Marine Air-Ground Task Forces.

- Enable the introduction of follow-on forces by securing staging areas ashore.
- Operate in rural and urban environments.
- Operate under nuclear, biological, and chemical warfare conditions.
- Withdraw rapidly at the conclusion of operations.
- Participate fully in the joint planning process and successfully integrate MAGTF operations with those of the joint force.

Elements

All MAGTFs are expeditionary by design and comprised of four core elements: a command element (CE), a ground combat element (GCE), an aviation combat

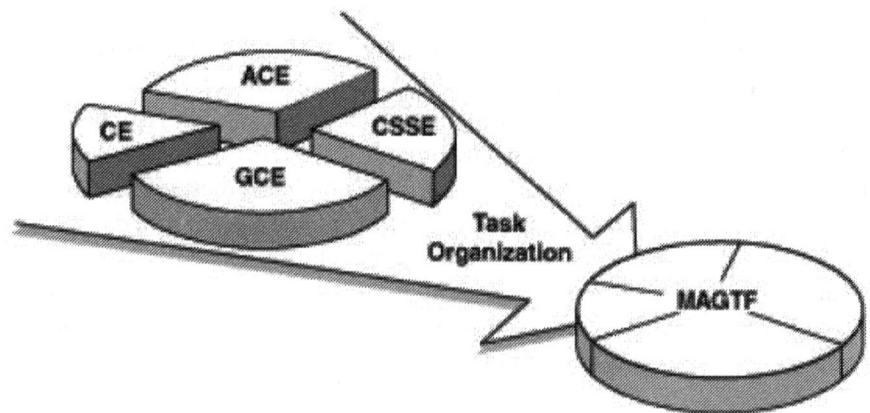

Figure 3-5. MAGTF Organization.

element (ACE), and a combat service support element (CSSE). See figure 3-5. The MAGTF's combat forces reside within these four elements. Although MAGTFs will differ because of mission and forces assigned, a standard procedure exists for organization, planning, and operations.

As a modular organization, the MAGTF is tailorable to each mission through task organization. This building block approach also makes reorganization a matter of routine. In addition to the Marine Corps units, MAGTFs may have attached forces from other Services and nations; e.g., naval construction force, multiple launch rocket system batteries, and armor brigades.

A key feature of Marine expeditionary organization is expandability. Crisis response requires the ability to expand the expeditionary force after its introduction in theater without sacrificing the continuity of operational capability. The MAGTF's modular structure lends itself to rapid expansion into a larger force as a situation demands by simply adding forces as needed to the core units of each existing element. This expandability includes expanding into a joint or multinational force because the MAGTF structure parallels the structure of a multidimensional joint force. Operation Restore Hope in Somalia is an example of the expandability of the MAGTF. This contingency began with the employment of a MEU (special operations capable) (MEU[SOC]) to seize the port and airport in Mogadishu, enabling the deployment of elements of I MEF via air and MPS, with the MEF eventually employing a brigade-sized force to provide security and humanitarian relief to the Somalis.

On missions where Marine forces are not deployed as part of a MAGTF (such as the Chemical/Biological Incident Response Force), Marine security forces, or forces assigned to battle forest fires, the Marine Corps component commander plays an increasingly important role in ensuring the Marine forces are properly equipped, trained, and employed. Marine forces deployed as a MAGTF are normally employed by the joint force commander as a MAGTF. As a task-organized force, the MAGTF's size and composition depend on the committed mission. If a MAGTF is deprived of a part of its combat forces, accomplishment of the mission for which it is tailored is jeopardized. However, on a day-to-day basis, the MAGTF may be tasked to conduct operations in support of another force and will identify capabilities; e.g., air sorties, beach and port operations, and civil affairs, excess to its mission requirements to the joint force commander that may be of use to other components of the joint force.

Command Element

The CE is the MAGTF headquarters. As with all other MAGTF elements, it is task-organized to provide the command and control capabilities necessary for effective planning, execution, and assessment of operations across the warfighting functions. Additionally, the CE can exercise command and control within a joint force from the sea or ashore and act as a core element around which a joint task force headquarters may be formed, provide interagency coordination for MOOTW, and conduct "reach back." The six warfighting functions are: command and control, intelligence, maneuver, fires, logistics, and force protection.

A CE may include additional command and control and intelligence capabilities from national assets and theater, force reconnaissance company assets, signals intelligence capabilities from the radio battalion, and a force fires coordination center. A CE can employ additional major subordinate commands such as the force field artillery headquarters, naval construction regiments, or Army maneuver or engineering units.

Ground Combat Element

The GCE is task-organized to conduct ground operations, project combat power, and contribute to battlespace dominance in support of the MAGTF's mission. It is formed around an infantry organization reinforced with artillery, reconnaissance, assault amphibian, tank, and engineer forces. The GCE can vary in size and composition from a rifle platoon to one or more Marine divisions. It is the only element that can seize and occupy terrain.

Aviation Combat Element

The ACE is task-organized to conduct air operations, project combat power, and contribute to battlespace dominance in support of the MAGTF's mission by performing some or all of the six functions of Marine aviation: antiair warfare, assault support, electronic warfare, offensive air support, air reconnaissance, and control of aircraft and missiles. It is formed around an aviation headquarters with air control agencies, aircraft squadrons or groups, and combat service support units. It can vary in size and composition from an aviation detachment of specifically required aircraft to one or more Marine aircraft wings. The ACE may be employed from ships or forward expeditionary land bases and can readily transition between sea bases and land bases without loss of capability. It has the capability of conducting command and control across the battlespace.

Combat Service Support Element

The CSSE is task-organized to provide all functions of tactical logistics necessary to support the continued readiness and sustainability of the MAGTF. The six functions of tactical logistics are: supply, maintenance, transportation, health services, engineering, and other services which include legal, exchange, food, disbursing, postal, billeting, religious, mortuary, and morale and recreation services. See MCWP 4, *Logistics*, for a detailed discussion. The CSSE is formed around a combat service support headquarters and may vary in size and composition from a support detachment to one or more force service support groups. The CSSE, operating from sea bases or from expeditionary bases established ashore, enables sustainment of forces, thus extending MAGTF's capabilities in time and space. It may be the main effort of the MAGTF during humanitarian assistance missions or selected phases of MPF operations.

Supporting Establishment

The supporting establishment is often referred to as the "fifth element of the MAGTF." It is vital to the success of Marine Corps forces conducting expeditionary operations. It recruits, trains, equips, and sustains Marines enabling them to conduct expeditionary operations in increasingly complex and dangerous environments. Bases and stations of the supporting establishment provide the training areas, ranges, and the modeling and simulation facilities necessary to prepare Marines and their units for combat. These posts of the Corps serve as staging and marshalling areas for deploying units and often are the CONUS end of a responsive replacement, supply, and new equipment pipeline into the AO. The Marines, sailors, and civilians of the supporting establishment are true partners with the Marines of the operating forces in accomplishing the mission.

Bases and stations of the supporting establishment also provide facilities and support to the families of deployed Marines, allowing Marines to concentrate fully on their demanding missions without undue concern for the welfare of their families.

Types

MAGTFs are integrated combined arms forces structured to accomplish specific missions. MAGTFs are generally categorized in the following four types.

Marine Expeditionary Force

The MEF is the Marine Corps' principal warfighting organization. It can conduct and sustain expeditionary operations in any geographic environment. MEFs are the sole standing MAGTFs; e.g., they exist in peacetime as well as wartime. Size and composition can vary greatly depending on the requirements of the mission. A MEF is normally commanded by a lieutenant general. It can be comprised of—

- A standing command element.
- A GCE of one or more divisions.
- An ACE of one or more aircraft wings.
- A CSSE of one or more force service support groups.

A MEF not only deploys and commands its own units, but also units from other MEFs, the Marine Corps Forces Reserve, other Services and nations, and the Special Operations Command. It typically deploys by echelon with 60 days of sustainment, but can extend operations with external support from other United States Services or through host-nation support agreements. The MEF commander and his staff can form the nucleus for a joint task force or functional component headquarters.

A MEF nominally consists of a permanent CE as well as a tailored Marine division, Marine aircraft wing, and Marine force service support group. Each MEF deploys a MEU(SOC) on a continuous basis to provide forward presence and crisis response capabilities to the combatant commanders. There are three standing MEFs:

- I MEF, based in southern California and Arizona, assigned to CINCUSPACOM.
- II MEF, based in North Carolina and South Carolina, assigned to CINCUSJFCOM.
- III MEF, based in Okinawa, mainland Japan, and Hawaii, assigned to CINCUSPACOM.

Marine Expeditionary Brigade

The Marine expeditionary brigade (MEB) is the "middle-weight" MAGTF. It is a crisis response force capable of forcible entry and enabling the introduction of follow-on forces. It can serve as part of a joint or multinational force and can provide the nucleus of a joint task force headquarters. It is unique in that it is the smallest MAGTF with a fully capable aviation element that performs all six functions of Marine aviation and is self-sustaining for 30 days. A MEB is capable of rapid deployment and employment deploying either by air, in combination with the MPS, or by amphibious shipping.

As a result, the MEB can conduct the full range of combat operations and may serve as the lead echelon of the MEF. The MEB is not a standing organization, but rather imbedded within the MEF. As a result, MEBs are task-organized for specific missions from within the assets of the MEF. The MEB conducts the mission or prepares for the subsequent arrival of the rest of the MEF or other joint or multinational forces. However, the deployment of a MEB does not necessarily mean that all the forces of the MEF will follow. Currently, the 1st, 2d, and 3d MEBs have been designated within I, II, and III MEF and are commanded by the deputy MEF commanders or other general officers. A MEB notionally consists of the following elements:

- A CE that may include additional assets such as command and control, force reconnaissance company, signals intelligence capabilities from the radio battalion, and engineering capabilities from the naval construction regiments. It can also control the forces of other Services and nations in missions ranging from combat in an urban area to disaster relief.
- A GCE composed of an infantry regiment reinforced with artillery, reconnaissance, engineer, light armored reconnaissance units, assault amphibian units, and other attachments as required.
- An ACE composed of a Marine aircraft group comprised of combat assault transport helicopter, utility and attack helicopters, vertical/short takeoff and landing fixed-wing attack aircraft, air refuelers/transport aircraft, and other detachments as required.
- A CSSE task-organized around a brigade service support group. This element has engineering, supply, transportation, landing support for beach, port and airfield delivery, medical, and maintenance capabilities.

The 4th MEB (AT) provides the unified combatant commanders with a rapidly deployable and sustainable specialized antiterrorism force to deter, detect, and defend against terrorist actions and conduct initial incident response to combat the threat of terrorism worldwide. See page 1-20.

Marine Expeditionary Unit (Special Operations Capable)

The MEU(SOC) is the standard forward-deployed Marine expeditionary organization. A forward-deployed MEU(SOC) provides an immediate sea-based response to meet forward presence and power projection requirements. A MEU(SOC) is commanded by a colonel and deploys with 15 days of supplies. It is normally comprised of—

- A CE that may include additional assets such as command and control, force reconnaissance company, and signals intelligence capabilities from the radio battalion.
- A GCE comprised of an infantry battalion reinforced with artillery, reconnaissance, engineer, tanks, light armored reconnaissance units, assault amphibian units, and other attachments as required.
- An ACE comprised of a combat assault transport helicopter squadron reinforced with utility and attack helicopters, vertical/short takeoff and landing fixed-wing attack aircraft, air refuelers/transport aircraft, and other detachments as required.
- A CSSE task-organized around a MEU service support group. This element has engineering, supply, transportation, landing support, medical, and maintenance capabilities.

A forward-deployed MEU(SOC) operates continuously in the Mediterranean Sea, the western Pacific Ocean, and the Indian Ocean or Arabian Gulf region. Embarked aboard a Navy amphibious squadron, the MEU(SOC) provides a combatant commander or other operational commander a quick, sea-based reaction force for a wide variety of missions such as limited forcible entry operations, noncombatant evacuations, raids, or disaster relief. In many cases, the MEU embarked on amphibious shipping may be the first United States force at the scene of a crisis and can enable the actions of larger follow-on forces. It can provide a visible and credible presence in potential trouble spots and can demonstrate the United States' willingness to protect its interests overseas. While the MEU(SOC) is not a special operations force per se, it can support special operations forces and execute certain maritime special operations missions. These include reconnaissance and surveillance; specialized demolitions; tactical recovery of aircraft and personnel; seizure/recovery of offshore energy facilities; seizure/recovery of selected personnel or material; visit, board, search, and seizure of vessels; and in extremis hostage recovery.

Prior to deployment, the MEU(SOC) undergoes an intensive 6-month training program focusing on its conventional and selected maritime special operations missions. Training culminates with a thorough evaluation and certification as "special operations capable." To receive this certification, a MEU must demonstrate competence across the entire spectrum of required capabilities, be able to plan and execute any assigned mission within 6 hours of notification, and conduct multiple missions simultaneously. Inherent capabilities of a MEU(SOC) are divided into four broad categories:

- Amphibious operations.
- Direct action operations.
- MOOTW.
- Supporting operations.

The complete list of capabilities subcategories for the MEU(SOC) is found in Marine Corps Order 3120.9A, *Policy for Marine Expeditionary Unit (Special Operations Capable) (MEU[SOC])*.

Special Purpose MAGTF

A special purpose MAGTF is a nonstanding MAGTF temporarily formed to conduct a specific mission for which a MEF or other unit is either inappropriate or unavailable. They are organized, trained, and equipped to conduct such a mission. Special purpose MAGTFs have been deployed for a wide variety of missions such as humanitarian relief and coalition training. Designation of a special purpose MAGTF is based on the mission it is assigned ("Special Purpose MAGTF Hurricane Relief"), the location in which it will operate ("Special Purpose MAGTF Somalia") or the name of the exercise in which it will participate ("Special Purpose MAGTF Unitas").

A special purpose MAGTF may be of any size—but normally no larger than a MEU—with narrowly focused capabilities required to accomplish a particular mission. It may be task-organized from nondeployed Marine Corps forces or formed on a contingency basis from a deployed MAGTF. Regimental-level headquarters often assume the role as a special MAGTF CE and may conduct training in anticipated mission skills prior to establishment. A special purpose MAGTF may be deployed using commercial shipping or aircraft, strategic airlift, amphibious shipping or organic Marine aviation.

An important type of special purpose MAGTF is the air contingency force (ACF). An ACF is an on-call, task-organized alert force that is maintained by all three MEFs. An ACF can deploy within 18 hours of notification. It can be dispatched virtually worldwide to respond to a rapidly developing crisis. The ACF is the MEF's force in readiness. It can deploy independently or in conjunction with amphibious forces, MPFs, or other expeditionary forces.

Because it can deploy so rapidly, readiness is paramount. Equipment and supplies intended for use as part of an ACF are identified and, where appropriate, stored and staged for immediate deployment. Personnel continuously focus on their tactical readiness. The ACF is airlifted to a secure airfield and carries its own initial sustainment.

The ACF is comprised of the same elements as any MAGTF although normally an ACF is a MEU-sized force. Due to the need to reduce to an absolute minimum the size and weight of an air deployed force, only those personnel and equipment needed to perform the function of each MAGTF element are included in the ACF.

MARINE LOGISTICS COMMAND

The Marine Corps component commander may establish a Marine Logistics Command (MLC) if he determines that the mission requires logistic support beyond what the CSSE can provide. The combatant command-level Marine Corps component commander may establish an MLC to fulfill his Service logistic responsibilities. The MLC is not a standing organization, but is task-organized to meet the operational support and sustainment requirements of the mission and is normally formed around a force service support group from another MEF. When formed, it provides logistic support to all Marine Corps forces in theater, and may provide limited support to other joint and multinational forces as directed by the combatant commander. The MLC provides operational logistics to Marine Corps forces as the Marine Corps component's logistics agency in theater. Operational-level logistics includes deployment, sustainment, resource prioritization and allocation, and requirements identification activities required to sustain the force in a campaign or major operation. These fundamental decisions concerning force deployment and sustainment are key for the MLC to provide successful logistical support.

The MLC provides the Marine Corps theater support structure necessary to facilitate reception, staging, onward movement, and integration (RSOI) of deploying Marine Corps forces. For more on RSOI, see chapter 4.

CHAPTER 4

Employment of Marine Corps Forces at the Operational Level

Contents	
Battlespace Organization	**4-3**
Area of Operations	4-4
Area of Influence	4-6
Area of Interest	4-7
Boundaries, Maneuver Control Measures, and Fire Support Coordinating Measures	4-7
Deployment	**4-10**
Force Deployment Planning and Execution	4-10
Predeployment Activities	4-14
Movement to and Activities at the Port of Embarkation	4-15
Movement to the Port of Debarkation	4-16
Reception, Staging, Onward Movement, and Integration	4-16
Deployment Forces	4-16
Employment	**4-18**
Planning Approach	4-19
Arranging Operations	4-20
Combat Power	4-21
Warfighting Functions	4-21
Information Operations	4-22
Redeployment	**4-24**
Expeditionary Operations in Support of Future Campaigns	**4-24**

"In forming the plan of a campaign, it is requisite to foresee everything the enemy may do, and be prepared with the necessary means to counteract it."

—Napoleon Bonaparte

Marine Corps expeditionary forces, comprised of a Marine Corps component command and one or more MAGTFs, are well-suited to perform full spectrum operations as part of a joint or multinational force at the operational level of war.

Uniquely equipped and trained to perform a variety of tactical actions from the sea, in the air, and on land, Marine Corps expeditionary forces are often the first forces to arrive in a crisis area. They then begin a series of actions to create conditions that meet the Nation's goals and strategic objectives. Marine commanders may have operational-level responsibilities that include making on-the-spot decisions that will set the course of the Nation's role in the crisis. They may also be responsible for providing the nucleus of a joint force headquarters. Regardless of the size of the forces involved or the scope of the military action, if Marine Corps expeditionary forces are operating to achieve a strategic objective, then they are being employed by the joint force commander at the operational level.

As stated in MCDP 1-2, *Campaigning*, the principal tool by which the joint force commander pursues the conditions that will achieve the strategic goal is the campaign. A campaign is a series of related military operations aimed at accomplishing a strategic or operational objective within a given time and space. Normally, there is only one campaign underway within a combatant commander's theater of war or theater of operations. This campaign is almost always a joint campaign and is under the command of the combatant commander or the commander of a subordinate joint task force. The joint force commander's campaign is normally a series of related operations, some of which may be conducted by a single Service, such as the Navy escorting foreign flag tankers through the Persian Gulf during the Iran-Iraq War in the 1980s. Other campaigns may consist of large operations conducted by joint and multinational forces such as the Marianas campaign in the Central Pacific in 1944. During this campaign, Marine, Navy, and Army forces conducted three major amphibious operations, seizing the islands of Saipan, Tinian, and Guam from the Japanese, while the Navy turned away the Japanese fleet in the Battle of the Philippine Sea.

The art of campaigning consists of deciding who, when, and where to fight and for what purpose. As a campaign is a series of operations or battles whose outcome lead to accomplishing the strategic goals, the commander must focus on these goals. Tactical and operational decisions must be made with these goals in mind. Ideally, operational commanders fight only when and where they want, maintaining the initiative and a high tempo of operations. They fight when they are stronger than the enemy and avoid battle when they are at a disadvantage. This allows them to shape the tactical actions to meet their strategic goals.

Campaigns are normally conducted by joint force commanders within a joint or combined operational framework. This framework provides for logical divisions of the battlespace to facilitate operations and clearly delineate Service responsibilities. Campaigning involves—

- Fighting a series of battles to accomplish the joint force commander's objectives.
- Defeating the enemy in depth by fighting the single battle.
- Synchronizing operations (not scripting activities but arranging activities in space and time.)
- Breaking the enemy apart over time. While destroying the enemy's will to resist in one attack is the ideal, the MAGTF may not have the ability to do this. A series of battles may be necessary.
- Knowing when and where to give battle and when to refuse battle.

BATTLESPACE ORGANIZATION

Understanding the joint battlespace at the operational level of war in which Marine Corps forces will operate is an important step in setting the conditions for their success. Marines must understand the relationship between the AO, area of interest, and area of influence. By analyzing his AO in terms of his area of influence and area of interest, a Marine commander determines whether his assigned AO is appropriate. This analysis may include the forces' capabilities to conduct actions across the warfighting functions.

Battlespace is the environment, factors, and conditions that must be understood to successfully apply combat power, protect the force, and accomplish the mission. This includes the air, land, sea, space, and enemy and friendly forces, infrastructure, weather, and terrain within the assigned AO and the commander's area of interest. Battlespace is conceptual—a higher commander does not assign it. Commanders determine their own battlespace based on their mission, the enemy, and their concept of operations and force protection. They use their experience and understanding of the situation and mission to visualize and adapt their battlespace as the situation or mission changes. The battlespace is not fixed in size or position. It varies over time, and depends on the environment, the commander's mission, and friendly and enemy actions. Battlespace is normally comprised of an AO, area of influence, and area of interest.

Joint force commanders may define additional operational areas or joint areas to assist in the coordination and execution of joint operations. The size of these areas and the types of forces used depend on the scope, nature, and projected duration of the operation. Combatant commanders and other joint force commanders use the following organization of the battlespace at the operational level of war.

Combatant commanders are assigned an area of responsibility in the Unified Command Plan. Areas of responsibility are also called theaters where the combatant commanders have the authority to plan and conduct operations. Within the theater, the combatant commanders may designate theaters of war, theaters of operation, combat zones, and a communications zone.

A theater of war is the area of air, land, and sea that is, or may become, directly involved in the conduct of the war. It may be defined by either the National Command Authorities or the combatant commander and normally does not encompass the combatant commander's entire area of responsibility.

A theater of operations is a subarea within a theater of war defined by the combatant commander required to conduct or support specific operations. Different theaters of operations within the same theater of war will normally be geographically separate and focused on different enemy forces. Theaters of operations are usually of significant size, allowing for operations over extended periods of time.

A combat zone is that area required by combat forces for the conduct of operations. The communications zone is the rear part of the theater of war or theater of operations (behind but contiguous to the combat zone) that contains lines of communications, establishments for supply or evacuation, and other agencies required for the immediate support and maintenance of the field forces. The Marine Corps component commander will normally focus his efforts to deploy, support, and sustain his forces, particularly the MAGTF, in the communications zone. He will normally locate his headquarters close to the joint force commander, who usually establishes his headquarters in the communications zone.

Area of Operations

An AO is an operational area defined by the joint force commander for land and naval forces. AOs do not typically encompass the entire operational area of the joint force commander, but should be large enough for the Marine Corps component commander and his subordinate units to accomplish their missions and protect their forces. The AO is the tangible area of battlespace and is the only area of battlespace that a commander is directly responsible for. AOs should also be large enough to allow commanders to employ their organic, assigned, and supporting systems to the limits of their capabilities. The commander must be able to command and control all the forces within his AO. He must be able to see the entire AO—this includes coverage of the AO with the full range of collections assets and sensors available to the Marine Corps component command and MAGTF, to include reconnaissance, electronic warfare aircraft, unmanned aerial vehicles, remote sensors, and radars. He must be able to control the events and coordinate his subordinates' actions. Finally, the commander must be able to strike and maneuver throughout the AO.

Marine Corps component commanders may assign AOs to their subordinates. These commanders typically subdivide some or all of their assigned AO to their subordinates. As operations unfold the Marine Corps component commander may adjust or reposition the MAGTF's AO and assume responsibility for what was the MAGTF's rear area. This allows the MAGTF to maintain momentum and have appropriate battlespace to continue the mission and protect the force.

The commander will use control measures to describe the AO, using the minimum necessary control measure to assign responsibility, promote unity of effort, and accomplish the unit's mission. Commanders should not assign areas of operations that are significantly greater than the unit's area of influence. A subordinate commander who is unable to directly influence his entire AO may have to request additional forces or assets that will allow him to extend his operational reach. Failing that he may have to—

- Request a change in mission or tasks.
- Request a reduction in the size of his AO.
- Revise his concept of operations by phasing operations in such a way that he only needs to directly influence portions of his AO.

The commander can choose to organize his AO so that his subordinates have contiguous or noncontiguous AOs. A contiguous AO is one where all subordinate commands' AOs share one or more common boundary while a noncontiguous AO is one where one or more subordinate AOs do not share a common boundary. Commands with contiguous AOs are normally within supporting distance of one another. The commander establishes contiguous AOs when—

- The AO is of limited size to accommodate the force.
- Political boundaries or enemy dispositions require concentration of force.
- There is a risk of being defeated in detail by enemy forces or the enemy situation is not clear.
- Key terrain is in close proximity to each other.

- Concentration of combat power along a single avenue of advance or movement corridor is required.

A noncontiguous AO is normally characterized by a 360-degree boundary. Because units with noncontiguous AOs must provide all-around security, they generally allow for less concentration of combat power along a single axis. There is additional risk associated with noncontiguous AOs in that units with noncontiguous AOs are normally out of supporting range of each other. The commander establishes noncontiguous AOs when—

- Limited friendly forces must occupy or control widely separated key terrain.
- Relative weakness of the enemy does not require that subordinate units remain within supporting range of each other.
- Dispersal of the enemy throughout the AO requires a corresponding dispersal of friendly units.

Operations against the enemy outside of noncontiguous subordinate commands' AOs are the responsibility of the common higher commander.

A joint operations area is determined and assigned by a combatant commander or a subordinate unified commander. It is the area of land, sea, and airspace in which a joint force commander (normally the commander of a joint task force) conducts military operations to accomplish a specific mission. In amphibious operations, the joint force commander may designate an amphibious objective area or an AO that contains the amphibious task force objectives.

Area of Influence

The area of influence is that geographical area wherein a commander is directly capable of influencing operations by maneuver or fire support systems normally under the commander's command or control. It is that portion of the battlespace that the commander can affect through the maneuver, fires, and other actions of his force. Its size is normally based on the limits of organic systems (fire support, aviation, mobility, and reconnaissance capabilities) and operational requirements identified within each of the warfighting functions. The area of influence normally reflects the extent of the force's operational reach. MAGTFs have significant areas of influence, employing Marine fixed-wing aviation, to extend the operational reach of Marine forces.

The commander considers his mission, forces, inherent warfighting functions requirements, and the AO to determine his area of influence. The area of influence is useful to the commander as a tool in assigning subordinate areas of operations and in focusing intelligence collection and information operations to shape the battlespace to facilitate future operations. Today's area of influence may be tomorrow's AO.

Area of Interest

The area of interest contains friendly and enemy forces, capabilities, infrastructure, and terrain that concern the commander. This area includes the area of influence and those areas that contain current or planned objectives or enemy forces that are capable of endangering mission accomplishment. The size of the area of interest normally exceeds the commander's operational reach.

While the area of interest includes the AO and area of influence, the area of interest may stretch far beyond the other parts of his battlespace. In analyzing the battlespace to determine his area of interest, the commander may pose the questions, "Where is the enemy and where are his friends?" and then "Where am I, and where are my friends?" The answers to these questions help identify the size, location, and activities that constitute the commander's area of interest. He may also consider critical information requirements such as critical terrain and infrastructure features and the ability of his intelligence and information assets to collect on these features when determining this area. The commander may request joint theater or national assets to help him understand the battlespace and collect intelligence throughout his area of interest.

Another key point to remember is that the area of interest may be noncontiguous. See figure 4-1 on page 4-8. For example, a forward-deployed MEF may have an area of interest back in the CONUS while the time-phased force and deployment list is being executed. It may also have areas of interest around airbases in other countries neighboring the MEF's AO. Using noncontiguous areas of interest conserves time and scarce collection assets. Assets will be allocated and time will be invested to provide the information required. Identifying noncontiguous points vice large generic areas is a technique that can conserve these valuable resources.

Boundaries, Maneuver Control Measures, and Fire Support Coordinating Measures

Joint force commanders and other commanders at the operational level of war often use boundaries, maneuver control measures, and fire support coordinating measures to control and coordinate the operations of their forces in the battlespace. These measures are usually employed to delineate areas of operation or other areas where components or subordinate commands will conduct their operations or to coordinate maneuver or fires between adjacent units. Each of these measures has specific and discrete purposes. Their use

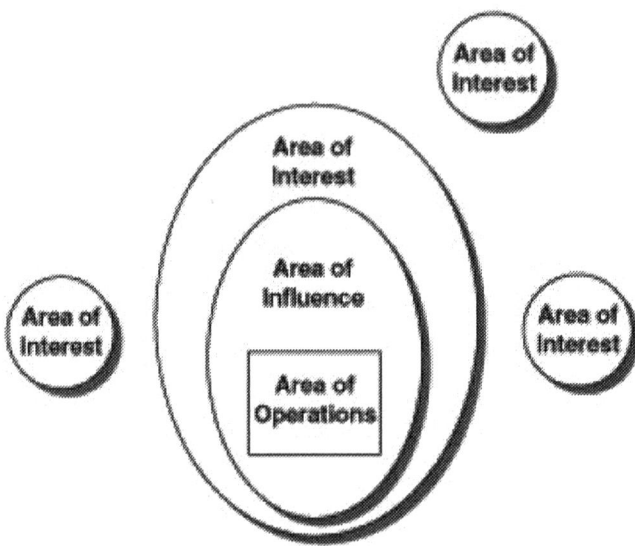

Figure 4-1. Noncontiguous Areas of Interest.

normally results in the units affected by them having to do something or refrain from doing something. Therefore, it is critical that Marine commanders use these measures as they were designed and that any modifications to the doctrinally based measure be clearly articulated to all affected higher, adjacent, and subordinate units.

The most commonly used of these measures is the boundary. A boundary is a line that delineates surface areas for the purpose of facilitating coordination and deconfliction of operations between adjacent units, formations or areas. They are used to define the forward, flank, and rear limits of an AO and when possible should be drawn along identifiable terrain to aid in recognition. See figure 4-2.

An axis and limit of advance are used at the operational and tactical levels of war. An axis of advance is a line of advance assigned for the purpose of control; often a road, network of roads, mobility corridor, or a designated series of locations, extending in the direction of the enemy. It provides subordinate commanders with a graphic representation of the commander's intent for their scheme of maneuver. Subordinate commanders are guided by it but may deviate from it when the situation dictates. A limit of advance is an easily recognizable terrain feature beyond which attacking elements will not advance. It is assigned by the commander to subordinate maneuver commanders to control their actions and form a limit of their advance.

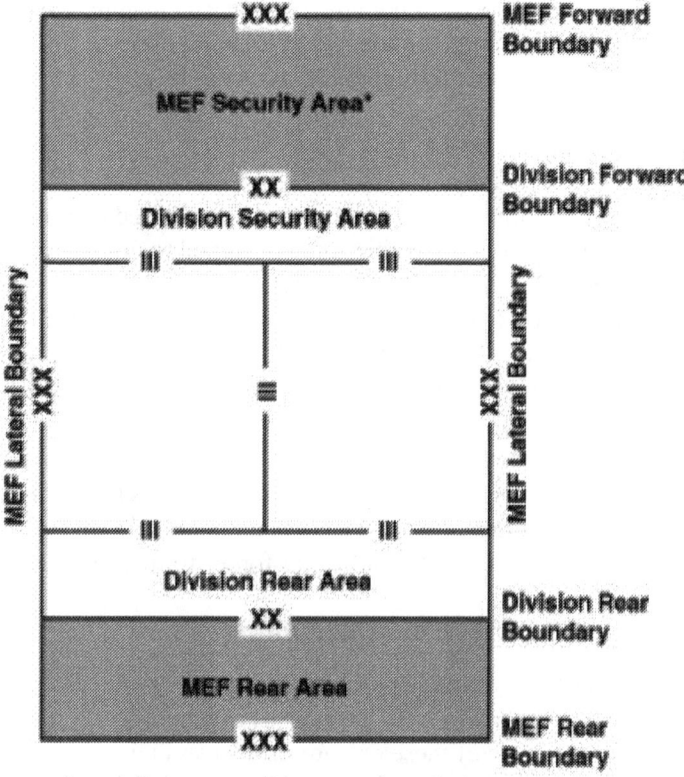

Figure 4-2. Unit Boundaries.

One of the most important and frequently misused fire support coordinating measures at the operational level is the fire support coordination line (FSCL). A FSCL is a permissive fire support coordinating measure used to facilitate timely attack of the enemy by air and surface-based fires. Supporting elements may engage targets beyond the FSCL without prior coordination with the establishing commander, provided the attack will not produce adverse effects on or to the rear of the FSCL or on forces operating beyond the FSCL. It is established and adjusted by the appropriate ground or amphibious force commander in consultation with superior, subordinate, supporting, and other affected commanders. The FSCL is not a boundary between aviation and ground forces and should not be used to delineate a de facto AO for aviation forces. It is located within the establishing commander's AO. Synchronization of operations on either side of the FSCL out to the forward boundary of the establishing unit is the responsibility of the establishing commander. When possible the FSCL should be drawn along readily identifiable terrain to aid in recognition.

DEPLOYMENT

Deployment is the relocation of forces and materiel to desired operational areas. The task-organized MAGTF is a flexible force designed to quickly and efficiently transition from peacetime to wartime and is designed for rapid deployment. MAGTFs are deployed as one of three basic types of forces; amphibious, MPF, and ACF. The MAGTF deploys by a variety of means—amphibious ships, strategic sealift, strategic airlift, and self-deploying aircraft. Deployment includes a range of activities from preparing to deploy, to transit, to arrival, right up to the beginning of employment. Deployment readiness is critical. Deployment planning is accomplished simultaneous to, and in conjunction with, employment planning. This planning entails all activities necessary to organize, train, educate, and equip forces for deployment.

Force Deployment Planning and Execution

The key to successful deployment is planning. While planning takes many forms, the Marine Corps engages in two basic categories of planning—force planning and operation planning. Both of these categories of planning impact on deployment. Force planning involves the activities necessary to recruit, train, educate, organize, equip, and provide forces. Operation planning involves the activities necessary to develop concepts, objectives, and tasks at all three levels of war—strategic, operational, and tactical. The Marine Corps integrates aspects of both categories of planning to ensure forces are prepared to deploy for immediate employment in a crisis. This integration is referred to as force deployment planning and execution. Proper force deployment planning and execution enables rapid deployment and responsive employment of forces.

Force deployment planning and execution includes the development of operational deployment procedures and the execution of those procedures. It is conducted with deliberate or crisis action planning in support of combatant commander requirements. Command elements at all echelons of the Marine Corps have an important part to play in force deployment planning and execution. The Marine Corps component commander and the MAGTF commander play critical roles.

The Marine Corps component commander is responsible for the development of supporting plans and time-phased force and deployment data, and exercises overall responsibility for the force deployment planning and execution for all Marine Corps forces assigned to a combatant commander. Normally, a subordinate joint command-level Marine Corps component commander focuses on those roles and responsibilities surrounding the employment and, to a lesser degree, redeployment of Marine Corps forces within the joint force. He has the same responsibilities as a combatant command-level Marine Corps component commander, but is primarily focused on crisis action planning. The combatant command-level Marine Corps component commander supports the subordinate joint command-level Marine component commander in executing his responsibilities.

The combatant command-level Marine Corps component staff works with the combatant commander's staff to prepare the operation plans and campaign plans for a theater. The component is the principal interface with the combatant commander for the development of required supporting plans to the operation plans. The component will task the MAGTF to accomplish detailed requirements determination; e.g., personnel, vehicles, sustainment, supplies required and associated shortfalls so the plan can be completed. The component commander is responsible for supporting those requirements. During deliberate planning, the MAGTF commander can direct the preparation of a supporting plan or operation orders. During the deliberate planning cycle, the Marine Corps component commander may prepare plans and operation orders to support the combatant commander's major theater war operation plans.

During deliberate planning, the Marine Corps component commander uses the MAGTF concept of operations as the focal point for his sustainment, support, and deployment planning. The component commander is ultimately responsible to the combatant commander for the deployment and subsequent employment of Marine Corps forces. He is assisted by the supporting establishment and other Marine Corps and external organizations; e.g., Marine Corps bases or Air Force airfield control units. Based on requirements identified to support the deployment and employment of the MAGTF, the component commander has the authority to enter into cross service and interservice support agreements, to submit for host-nation support, and to commit to contracts with civilian support agencies to support Marine Corps forces deployment.

The Marine Corps component commander determines the Marine Corps forces to be used in support of the combatant commander's requirements and advises him on the proper employment of the force. The component commander will publish joint operation planning and execution system planning guidance and a time-phased force and deployment data letter of instruction to aid the MAGTF in plans development. The component commander reviews the concepts of deployment and employment developed by the MAGTF commander and consolidates the time-phased force and deployment data inputs with those the component staff developed in support of the MAGTF. He then ensures the information is integrated into the combatant commander's operation plan and entered into the strategic deployment system. Planning considerations include the commander's intent for deployment and employment, the time-phasing of the forces and their sustainment, and the forces' closure requirements and responsibilities.

Sustainment planning is important as forces rapidly lose capabilities without adequate and dependable sustainment, particularly in an expeditionary environment. The component commander directs the deployment of the MAGTF and sources both the sustainment to support the deployment and to support the needs of the MAGTF during employment. The MAGTF and its major subordinate commanders identify any sustainment shortfalls not previously recognized by the component commander or the theater. The component commander coordinates sourcing sustainment to fill such critical shortfalls and ensures the earliest deployment of those assets. See chapter 5.

The Marine Corps component conducts both deliberate and crisis action planning. Deliberate planning is the joint process used primarily during peacetime and in prescribed cycles to develop joint operation plans; e.g., operation plans and contingency plans. Crisis action planning is the process used in response to an imminent crisis to develop joint operation plans or orders. During deliberate or crisis action planning, the component accomplishes the following tasks in support of deployment:

- Determines the force required to meet the combatant commander's requirement.
- Ensures that command relationships that maximize MAGTF capabilities are established.
- Provides planning guidance/letters of instruction to the MAGTF, supporting establishment, and Marine Corps Forces Reserve, as needed to guide deployment and sustainment planning.
- Registers all force and sustainment requirements in the time-phased force and deployment data.
- Fills or sources sustainment requirements.
- Fills or sources force and sustainment shortfalls.
- Monitors the task organization and deployment preparations of the force and sustainment.

- Coordinates CONUS and overseas bases and stations deployment support to the MAGTF.
- Arranges for interservice and host-nation support agreements.
- Integrates deployment planning information with the combatant commander's plan for deployment and employment of the joint force.
- Develops necessary supporting plans and orders.

The MAGTF commander accomplishes the detailed planning needed to prepare the forces to deploy and to employ. The MAGTF commander reviews the component commander's planning guidance and publishes guidance to his major subordinate commanders on planning for deployment and employment of the MAGTF. The major subordinate commanders follow this guidance to prepare their planning data, requirements, and operation orders for deployment and employment. This guidance is critical as it focuses their planning and ensures integrated planning. The major subordinate commanders integrate their planning efforts with the MAGTF commander's planning efforts. The MAGTF commander integrates the MAGTF planning effort with the component commander's efforts. Each commander and staff have specific tasks in planning and execution for which they are responsible, but there is also some overlap. This overlap is key to ensuring all aspects of the planning effort are addressed; e.g., requirements and shortfalls in personnel and supplies or phasing. During crisis action planning, the component, MAGTF, and major subordinate commands' staffs may conduct concurrent planning with the six steps of the Marine Corps Planning Process (MCPP) occurring near simultaneously. For more information on the MCPP, see MCWP 5-1, *Marine Corps Planning Process*.

Commanders may use reverse planning when the tactical employment of the forces dictate the operational and strategic deployment of the forces and their sustainment into a theater of operations. By envisioning the intended employment of the force, the MAGTF commander can better determine how the plan should reflect the arrival time of a specific unit and their support's deployment. In other cases, reverse planning will only be used as a means of planning to the final assembly area after arrival in the theater of operations.

During deliberate or crisis action planning, the Marine Corps component commander, MAGTF commander, and their staffs accomplish the following tasks in support of deployment:

- Determine the task organization of the force required to accomplish the mission.
- Develop a concept of deployment for the Marine Corps forces that supports force closure and employment.

- Provide planning guidance/letters of instruction to the major subordinate commanders as needed to guide their deployment and sustainment planning.
- Determine force shortfalls.
- Determine sustainment requirements, shortfalls, and phasing.
- Determine host nation or other service support requirements.
- Direct the preparation of embarkation and load plans.
- Determine the phasing of the assigned forces.
- Register all requirements in the time-phased force and deployment data.
- Develop a supporting plan or order as required.
- Submit all requirements to the component commander.
- Monitor execution of the deployment plan.

The deployment plan is developed with the employment plan. To facilitate effective and efficient execution of the deployment, the MAGTF identifies the requirements for support in—

- Personnel.
- Movement.
- Medical.
- Supply.
- Services.
- Maintenance of nondeploying equipment left in garrison.

Deployment is conducted in four general phases. The phases are predeployment activities, movement to and activities at the port of embarkation, movement to the port of debarkation, and RSOI.

Predeployment Activities

Predeployment includes two types of activities. The first is the daily activities that are part of the units' ongoing training, education, exercising, and health and comfort efforts. The unit's readiness for deployment and employment as a task-organized combat force is central to its expeditionary capability. Units address readiness in the areas of:

- Unit training.
- Personal readiness.
- Medical and dental.

- Supply and equipment.
- Administrative and legal.

Preparedness is a key to deployment. Leaders at all levels, from the component commander to the last fire team leader in a squad, must understand their responsibilities for organizational readiness. Readiness in each of the above areas is essential if a Marine Corps force is to be prepared to deploy at any time, anywhere in the world.

The second type of predeployment activity includes those preparations made immediately prior to a deployment. These activities include packing, marking, weighing, staging, and manifesting personnel and equipment. The mode of transportation, airlift or sealift, determines the time frame for conducting these activities. A pivotal consideration is whether or not the mode of transportation is to be administratively or combat loaded. Method of entry into the theater also dictates the loading consideration and is critical to preparation activities. To ensure the highest state of readiness, the MAGTF maintains designated quantities of equipment and supplies in a constant state of predeployment preparedness. These assets are stored and staged already packed, marked, weighed, and manifested for immediate deployment by air or sea. The ACE must be ready to self deploy its aircraft to the AO on short notice, so assets to support this movement must be readily available for deployment on support aircraft. The equipment and supplies necessary to operate expeditionary airfields or to operate from existing airfields in austere locations will usually be greater than those needed to operate from more developed sites.

Movement to and Activities at the Port of Embarkation

Movement to and activities at the port of embarkation and aerial port of embarkation is the second phase in deployment. The MAGTF supervises these activities. The MAGTF publishes guidance and procedures necessary to move to the port of embarkation and to prepare for movement into the theater of operations. Based on the requirements of the deployment and with the assistance of the supporting establishment, the MAGTF stands up the necessary organizations to facilitate this phase. They include the force movement control center, logistic and movement control center, and unit movement control centers to conduct the preparation and movement of the forces to the port of embarkation.

The force movement control center, as the MAGTF's senior movement agency, issues movement policy and guidance, and supervises the movement activities on behalf of the MAGTF commander. Personnel from all major subordinate commands man the logistic movement control center, located at the force service support group. It orchestrates the staging and movement of all forces and equipment to the appropriate ports of embarkation. The unit movement control center coordinates the preparation, staging, and movement of the subordinate units' personnel and equipment as directed and supported by the logistic movement control center. The force flows to the port of embarkation in accordance with the RSOI plan. It normally remains at the port of embarkation for a brief time, and then commences the next phase of deployment.

Movement to the Port of Debarkation

Movement to the port of debarkation is the third phase of deployment and is normally strategic movement. The forces will flow to the theater of operations via sealift or airlift according to the deployment plan in order to expedite force closure. In many cases, Marine squadrons will self-deploy to the theater and establish forward operating bases. When flowing via airlift, little integration can be accomplished prior to arrival at the port of debarkation. However, if the force is moving via sealift, some force integration can be accomplished. As sealift movement time is measured in days and weeks, the forces embarked on board can train and can further prepare their equipment for immediate use once in theater. This is especially critical in an amphibious operation where forcible entry is required.

Reception, Staging, Onward Movement, and Integration

RSOI is the fourth phase of deployment and includes those basic activities involving the personnel, equipment, and supplies of the MAGTF and their integration into warfighting capabilities. The Marine Corps integrates its combat power as soon as forces begin to arrive in the theater of operations. In amphibious operations, the integration of the MAGTF's combat power can be accomplished aboard ship prior to the assault. In MPF operations, the development of warfighting capabilities is accomplished ashore as soon as the off-loaded equipment and supplies are combined with the fly-in echelon. It is the RSOI phase that completes force closure and enables the MAGTF to transition into employment. The Marine Logistics Command, if established, may be used to conduct RSOI, as well as provide component-level logistic support to Marine forces in theater.

Deployment Forces

Each of the three deployment methods—amphibious, MPF, and ACF—influence employment and deployment planning and can influence the MAGTF's focus on readiness.

Amphibious forces remain the key option for the MAGTF to deploy forces into a theater where forcible entry is a likely requirement. They are also the primary means by which the Marine Corps provides a credible forward presence throughout the world. Amphibious forces allow the MAGTF to be indefinitely sustained from a protected area at sea, increase force endurance by allowing force reconstitution in a protected area, increase flexibility for force entry into the theater of operations, and offer a credible deterrent to hostilities. The MAGTF must have personnel, equipment, and supplies prepared at all times for deployment as an amphibious force and consider the implications on employment and deployment planning. In particular, the MAGTF commander considers the—

- Advantages/disadvantages of combat versus administrative loading of the ships to support the mission.
- Potential degradation of specific combat skills that cannot be exercised while at sea for extended periods of time.
- Advantages/disadvantages of deploying required sustainment with the force.
- Potential degradation of equipment readiness due to the lack of use for extended periods of time.
- Inherent flexibility in off-load methods available; e.g. boat, helicopter, air-cushioned landing craft, amphibious assault vehicle, and pier side.
- Advantages/disadvantages of conducting a rehearsal of the tactical tasks assigned when time allows during the at sea transit period.
- Advantage gained by not requiring a benign port or airfield for preparing for combat or assembling forces and their equipment.

The MPF allows the MAGTF to rapidly deploy to nearly all areas of the world. The MPF can deploy independently to a secure arrival/assembly area or it can deploy with an amphibious force that secures the arrival/assembly area. This flexibility provides the MAGTF commander with several employment options, which may influence his deployment planning and execution. The flexibility of the MPF not only allowed for the rapid deployment of 7th MEB during the early days of Operation Desert Shield (1990–1991), it also sustained the Army's XVIII Airborne Corps as it deployed into theater. The MAGTF considers the implications MPF operations have on employment and deployment planning. In particular, the MAGTF commander considers the—

- Advantages for rapid deployment gained because the majority of the MAGTF's equipment, vehicles, and supplies are pre-positioned.
- Speed with which personnel can be deployed via airlift.

- Requirement for large staging areas for the integration of personnel and equipment.
- Advantages in being able to incrementally off-load pier side or across a beach.
- Flexibility in being able to incrementally build the size of a force and its combat capability based on the mission requirements.
- Advantage in sustainability that the MPF provides to the MAGTF.

The ACF is a MAGTF that specializes in crisis response and can be used to reinforce MEUs, provide the lead echelon of a MEB, link up with the MPF, and support humanitarian assistance or disaster relief operations. The MEF commander considers the implications ACF operations have on employment and deployment planning. In particular, he considers—

- Advantages of the rapid deployment capability of the ACF.
- Impact the requirement for rapid deployment has on readiness.
- The requirement to prepare all of the ACF's organic equipment for deployment with personnel.
- The limited self-sustainment capability until link up with other forces, such as the MPF.
- The requirement for augmenting or sustaining the ACF for an extended period of time.
- The potential movement and resupply difficulties the ACF will encounter if the arrival airfield is a great distance from the AO.
- The requirement to self-deploy the ACE of the ACF.

To rapidly and efficiently deploy by any of these three methods requires all personnel, supplies, and equipment to be prepared. A detailed plan, which includes the procedures for executing the deployment, should be developed and exercised periodically.

EMPLOYMENT

Deployment of expeditionary forces to a foreign setting and their visible, credible presence offshore may be sufficient to accomplish national objectives or deter further crisis. However, when presence has not achieved friendly intentions, the actual employment of forces may be required to achieve military objectives and political goals. Employment is the use of Marine Corps forces to conduct operations to achieve the objectives of the joint force commander. This employment is comprised of the operational use of Marine Corps forces by the Marine Corps component commander or functional component commander and the tactical use of the MAGTF within the AO to attain military objectives. Employment includes both combat operations as well as MOOTW.

During employment, it is not unusual for a MAGTF to conduct offensive and defensive operations simultaneously. Early in Operation Desert Storm, I MEF aviation forces attacked Iraqi forces and installations in Kuwait while still maintaining a defensive combat air patrol over friendly forces in Saudi Arabia. Similarly, while I MEF ground units were attacking enemy critical vulnerabilities by use of artillery raids, adjacent light armored vehicle units were protecting friendly vulnerabilities by conducting counterreconnaissance missions.

Simultaneous combined arms operations, like those discussed above, and the need to operate across the entire range of military operations require MAGTFs to operate in an environment of mission depth. Marines have recently experienced mission depth in Somalia and Haiti with Marines on one city block providing humanitarian assistance, while on the next block dealing with civil disturbance, and on yet another block fully engaged in armed combat.

Planning Approach

Planning for employment is based on the factors of METT-T that provide a framework for assessing the elements of the battlespace (infrastructure, threat, friendly and noncombatants) to determine employment requirements. The mission establishes the context, relevancy, depth, and objectives of friendly operations. The enemy (or the threat—famine, political unrest, natural disaster) must be considered in the context of the environment and the elements that comprise it, such as the populace, governmental institutions, industry, and media. Terrain and weather determine the physical parameters of the battlespace—where the force interacts with the enemy. Troops and support available are the friendly resources and capabilities to be applied against the enemy or threat to accomplish the mission and achieve the desired effects. Time helps determine the scope of the battlespace—its depth and the extent and duration of operations. METT-T is used by the commander to prepare his forces to accomplish the mission, defeat the enemy, and operate effectively in the surrounding environment.

This approach to employment planning promotes a unifying effort within the force. Just as the enemy cannot be viewed in isolation, each MAGTF element has its own unique characteristics, capabilities, and in some cases, AO. Yet the efforts of the elements, when focused toward a common purpose, synergistically create combat power far greater than that possible through their independent employment.

Arranging Operations

The principal tool by which the operational level commander pursues the conditions that will achieve the strategic goal is the campaign. The MAGTF commander arranges his operations to support the joint force commander's campaign and fulfill the joint force commander's operational goals. The MAGTF commander's focus must be on achieving these operational goals. The MAGTF commander and his staff must understand the following key issues when arranging operation—

- The higher headquarters commander's intent and end state.
- The MAGTF's wider interaction with the joint force and Marine Corps component than that of its major subordinate commands.
- How the MAGTF gets to the AO particularly through force deployment planning and execution and RSOI.
- Theater logistics, specifically how to sustain the force over time and in depth.

Ideally, Marine commanders want to attack the enemy simultaneously across the width and breath of the battlespace to overwhelm the enemy's capabilities and reduce his options to counter. The simultaneous attacks throughout Panama by Army and Marine Corps forces during Operation Just Cause (1989) are examples of the devastating effect on the enemy's ability to command and control his forces. The scope, complexity, and duration of military operations during the campaign may require Marine Corps forces to phase their operations. A phase represents a period where a large portion of the forces are involved in similar or mutually supporting activities or there is a distinct type of operation. A transition from one phase to another may indicate a shift in emphasis, the main effort or the objective. The point where one phase stops and another begins is often difficult to define in absolute terms. During planning, commanders establish conditions for transitioning from one phase to another. The commander bases these conditions on measurable effects on the enemy, friendly forces, and the battlespace resulting from his and the enemy's actions. He must determine these effects and his procedures for collecting information to make his assessment of the conditions early in the planning process. During execution, commanders adjust the phases to exploit opportunities presented by the enemy or to react to unforeseen situations.

Phasing assists commanders in thinking through the entire operation and defining requirements in terms of forces, resources, time, and desired effects. A significant benefit of phasing is that it assists commanders in achieving major objectives by planning manageable subordinate operations to gain progressive advantages against the enemy. A significant disadvantage of phasing is the possible loss of tempo or momentum when transitioning between phases. This can create a period of vulnerability for the force. The enemy can take advantage of additional time and lack of pressure during any loss of tempo to consider his options and take more calculated actions to counter friendly operations. Another disadvantage of phasing is that it may create a discernable and predictable pattern that the enemy can use to disrupt or defeat friendly operations. The commander must be aware of the advantages and disadvantages of phasing. The commander can reduce the risks associated with transitioning between phases by overlapping phases and by anticipating potential problems.

Branches are options built into the basic plan. They may require shifting priorities, changing task organizations, and command relationships or change the very nature of the operation itself. Branches add flexibility to plans by anticipating situations that could alter the basic plan. The requirement for a branch plan may surface during the course of action war game step or as a result of the execution drills and rehearsals conducted during the transition of the plan. Such situations could arise from enemy action, availability of friendly capabilities or resources or even a change in the weather or season within the operational area. Sequels are subsequent operations based on the possible outcomes of the current operation—victory, defeat, or stalemate. Well-developed sequels help maintain momentum and facilitate the exploitation of fleeting opportunities.

Combat Power

MCDP 1, *Warfighting*, describes combat power as the total destructive force we can bring to bear on our enemy at a given time. MCDP 1-3, *Tactics*, describes it as a unique product of a variety of physical, moral, and mental factors. Commanders conceptually combine the elements of combat power—maneuver, fires, leadership, force protection, and information—to create overwhelming effects that lead to the defeat of the enemy. By combining and synchronizing these effects at the decisive place and time, the commander can convert the potential of his forces, resources, and opportunities into combat power. He then uses his combat power to reduce that of the enemy by destroying or disrupting the enemy's capability and will to resist. Combat power is generated through integrated planning and synchronization of the warfighting functions.

Warfighting Functions

The warfighting functions encompass all military activities performed in the battlespace. Warfighting functions are a grouping of like activities into major functional areas to aid in planning and execution of operations. These functional areas currently include command and control, maneuver, fires, intelligence, logistics, and force protection. The key advantage of using warfighting functions is that they allow the commander and his planners to look at all aspects of the battlespace and not leave anything to chance if it is within their capability to coordinate, control, influence, and synchronize them. By synchronizing the warfighting functions, the commander can increase the force's combat power, mass effects on the enemy, and aid in the assessment of the success of the operation. As stated in MCDP 1-2, maximum impact is obtained when all warfighting functions are synchronized to accomplish the desired objective within the shortest time possible and with minimum casualties.

Planners consider and integrate the warfighting functions when analyzing how to accomplish the mission. Integrating the warfighting functions helps to achieve focus and unity of effort. They provide a method for planners to think in terms of how each function supports the accomplishment of the mission. Critical to this approach to planning is the coordination of activities not only within each warfighting function but also among all the warfighting functions. By using warfighting functions as the integration elements, planners ensure all functions are focused toward a single purpose. The warfighting functions apply equally to conventional operations and other types of operations such as MOOTW and information operations. See appendix A.

Information Operations

Information operations (IO) include all actions taken to affect enemy information and information systems while defending friendly information and information systems. IO consist of the following distinct elements that must be employed together in an integrated strategy to succeed:

- Operations security.
- Psychological operations.
- Military deception.
- Electronic warfare.
- Physical destruction.
- Computer network operations.

IO related activities include public affairs and civil military operations that help shape civilian perceptions of MAGTF operations at home and in the AO.

IO are conducted across the range of military operations and at every level of war. They are well-suited to support expeditionary operations since they can project United States influence and can be tailored to provide measured effects in a specific mission or situation. They are scalable, allowing the commander to increase or decrease the level of intensity to reflect a changing situation. IO must be closely coordinated with those of the joint force commander to ensure unity of effort and to avoid undermining the effects desired by higher headquarters.

The primary focus of MAGTF IO is at the operational and tactical levels of war. Offensive IO are oriented against command and control targets, disrupting or denying an enemy's use of information and information systems to achieve the commander's objectives. The MAGTF relies primarily on electronic warfare and physical destruction to attack command and control, intelligence, and other critical information-based targets the enemy needs to conduct operations. The MAGTF can also employ deception operations to deceive the enemy commander's intelligence collection, analysis, and dissemination systems. MAGTF defensive IO protect information and information systems the MAGTF commander requires to plan and execute operations.

In some environments, IO capitalizes on the growing sophistication, connectivity, and reliance on information technology and focuses on the vulnerabilities and opportunities presented by the increasing dependence of the United States and its adversaries on information and information systems.

In other situations, IO may mean employing decidedly low-tech means to facilitate civil-military operations, influence selected target audiences or support military deception. Whatever the nature of the conflict, IO targets information or information systems to affect the information-based decisionmaking process. IO may, in fact, have its greatest impact as a deterrent in peace and during the initial stages of crisis. IO may help deter adversaries from initiating actions detrimental to the United States. At every echelon of command and all levels of warfare, some form of IO is likely to be a critical tool in achieving the objectives of the commander.

IO are a combination of offensive, defensive, and informational capabilities that are integrated and concurrently planned. Effective IO planning requires a framework that focuses the staff, ensuring a plan that supports the commander's concept of operations by integrating the elements of IO into a coherent, synchronized plan. In a sense, defensive IO is the shield to protect our own systems and decision processes and offensive IO is the sword used against the adversary. IO includes perception management, or those actions taken to influence selected groups and decisionmakers. Perception management combines informational activities, truth projection, operations security, cover and military deception, and psychological operations. It encompasses all actions taken to convey (or deny) selected information to an audience. It may be a key contributor to battlespace shaping efforts.

REDEPLOYMENT

Redeployment is the transfer of a unit, an individual or supplies deployed in one area to another area; to another location within the area for the purpose of further employment. It may also include movement back to home station. Redeployment is planned and executed based on mission requirements and with the same level of detail as was the original deployment. The same procedures and systems used to deploy the force are also used to support redeployment. Many of the same issues experienced during deployment will require similar detailed consideration during redeployment. Redeployment planning is conducted concurrently with deployment planning and continues until redeployment is completed.

A key difference between the Marine Corps and other Services is that during redeployment Marines focus on reestablishing the force's readiness for subsequent deployment and employment rather than merely returning to their home base. This ability to "re-cock" provides the joint force commander and the National Command Authorities with adaptable forces that can rapidly reload on amphibious or maritime pre-positioning shipping and conduct new missions in or out of theater. Operation Sea Angel in 1991 is an example of how Marine Corps forces undertook varied missions even while redeploying. Leaving the Persian Gulf after the successful completion of Operation Desert Storm, Marine Corps amphibious forces in transit to the United States were rapidly employed to conduct humanitarian assistance operations in flood ravaged Bangladesh.

As Marine Corps forces begin redeployment it may be necessary to reconstitute all or part of the force. Reconstitution entails those actions that commanders plan and take to restore units after operations to a desired level of combat readiness. This process may require reorganization of units that have sustained losses in personnel and equipment and, in cases of severe losses and damage, regeneration of units with replacements and new equipment. Reconstitution aims to restore the readiness of the force as quickly as possible. Units begin redeployment activities as soon as the operational employment has ceased for that unit. As redeployment begins, units also commence activities such as training and maintenance to restore and maintain combat readiness for its next mission.

EXPEDITIONARY OPERATIONS IN SUPPORT OF FUTURE CAMPAIGNS

Expeditionary maneuver warfare provides an axis of advance for future developments in how Marine Corps forces will conduct expeditionary operations. Technology—advanced aircraft, advanced amphibious assault vehicles, and enhanced command and control systems—with expeditionary maneuver warfare will improve the strategic agility, operational reach, support and sustainment, and joint and multinational interoperability of Marine Corps forces. Expeditionary operations will be conducted from sea bases—ships and expeditionary bases beyond the reach of the enemy's defenses—with forces rapidly maneuvering from the sea directly to objectives far inland. The operating forces will collaborate closely with the supporting establishment to deploy even more rapidly, organizing "on the fly"—tailoring and refining the mix of capabilities the scalable force brings to the crisis. Forcible entry will continue to be the focus of the Marine Corps as it employs new supporting concepts to overcome the challenges posed by the littoral area access denial steps taken by future adversaries.

The Marine Corps will capitalize on innovation, experimentation, and technology, as well as changes to doctrine, organization, training, and education, to provide competent and effective forces and capabilities to the joint force commander in future campaigns.

CHAPTER 5
Logistics in Marine Corps Operations

Contents

Levels	5-3
Strategic Logistics	5-4
Operational Logistics	5-4
Tactical Logistics	5-5
Functions	5-5
Supply	5-6
Maintenance	5-6
Transportation	5-7
General Engineering	5-7
Health Services	5-7
Services	5-7
Joint and Multinational Operations	5-8
Strategic Logistics Support	5-10
Strategic Mobility	5-10
United States Transportation Command	5-11
Department of Transportation	5-12
Defense Logistics Agency	5-12
Operational Logistics Support	5-12
Marine Corps Component	5-12
Marine Logistics Command	5-13
Reconstitution	5-14
Command and Control	5-15
Logistics Support to MAGTF Operations	5-16
Marine Expeditionary Force	5-17
Marine Expeditionary Brigade	5-18
Marine Expeditionary Unit (Special Operations Capable)	5-18
Special Purpose MAGTF	5-18
Air Contingency Force	5-19
Maritime Pre-positioning Forces	5-19
Aviation Logistics Support Ship	5-19
Norway Geopre-positioning Program	5-19
War Reserve Materiel Support	5-20
Tactical Level Command and Control	5-20
Supporting Establishment	5-20
Marine Corps Combat Development Command	5-21
Marine Corps Materiel Command	5-21
Marine Corps Bases, Stations, and Reserve Support Centers	5-22
Department of the Navy Agencies	5-22
Future Sustainment Operations	5-23

> *"Logistics is a key component of any and every operation of war. While it does not determine the shape that operations take, it sets limits that restrict the options available to commanders. Thus, the more flexible and far-reaching the logistics, the greater the possibility for bold, decisive, and imaginative action."*
> —MCDP 4, *Logistics*

Logistics is the science of planning and carrying out the movement and maintenance of forces. Logistics encompasses all actions required to move and maintain forces. This includes the acquisition and positioning of resources as well as the delivery of those resources to the forces. The terms logistics and combat service support are often used interchangeably, but there is a distinction. Logistics is the larger of the two concepts. Combat service support is the activity that actually provides essential services and supplies to operating forces in theater. Since most of the delivery of resources occurs at the tactical level of war, combat service support has been considered to be essentially the same as tactical logistics. Specific logistic needs are tailored to meet the conditions and the level of crisis or conflict under which a military force operates.

United States Code, Title 10, assigns each Service responsibility for organizing, training, and equipping forces for employment in the national interest. The Commandant is responsible for Marine Corps logistics. The Commandant ensures that Marine Corps forces under the command of a combatant commander or the OPCON of a unified, sub-unified or joint task force commander are trained, equipped, and prepared logistically to undertake assigned missions.

Logistics is a fundamental element of MAGTF expeditionary operations. Marine expeditionary forces provide self-contained and self-sustained forces that have the logistics to accomplish the mission from individual equipment to expeditionary airfields and medical treatment facilities. These forces are structured to meet a wide range of contingency operations and possess the logistic capabilities needed to initiate an operation, sustain forces, and reconstitute for follow-on missions. The Marine Corps' logistic responsibilities include—

- Preparing forces and establishing reserves of equipment and supplies for the effective prosecution of war.
- Planning for the expansion of peacetime forces to meet the needs of war.
- Preparing budgets for submission through the Department of the Navy based on input from Marine operating forces and Fleet Marine Force commanders assigned to unified commands.
- Conducting research and development and recommending procurement of weapons, equipment, and supplies essential to the fulfillment of the combatant mission assigned to the Marine Corps.

- Developing, garrisoning, supplying, equipping, and maintaining bases and other installations.
- Providing administrative and logistics support for all Marine Corps forces and bases.
- Ensuring that supported unified commanders are advised of significant changes in Marine Corps logistic support, including base adjustments that would impact plans and programs.

The Commandant, as a member of the Joint Chiefs of Staff, ensures that the Marine Corps—

- Prepares integrated logistic plans, which include assignment of logistic responsibilities.
- Prepares integrated plans for military mobilization.
- Reviews major personnel, materiel, and logistic requirements in relation to strategic and logistic plans.
- Reviews the plans and programs of commanders of unified and specified commands to determine their adequacy, feasibility, and suitability for the performance of assigned missions.

The Commandant vests in Marine Corps commanders the responsibility and authority to ensure that their commands are logistically ready for employment and that logistic operations are efficient and effective. This authority is exercised through administrative command channels for routine matters of logistic readiness and service planning. Designated commanders (usually at the Marine Corps component and/or MAGTF level) are also under the OPCON of unified, subordinate unified, and/or joint task force commanders for planning and conducting specified operations. Marine Corps component commanders, MAGTF commanders, and their subordinate commanders exercise the appropriate logistic responsibilities and authority derived from the joint force commander of a specified operation. Operational assignments do not preclude Service administrative command responsibilities and obligations. Commanders in the operating forces, supporting establishment, and the Marine Corps Reserve delegate authority for logistic matters to designated subordinates.

LEVELS

The strategic, operational, and tactical levels of logistics function as a coordinated whole, rather than as separate entities. Although the Marine Corps generally focuses on the tactical level of logistics, it is imperative that all Marines understand the interaction of all three logistic levels. These levels interconnect like sections of a pipeline, tying together logistics at the strategic, operational, and tactical levels.

Strategic Logistics

Strategic logistics supports organizing, training, and equipping the forces needed to further the national interest. It links the national economic base (people, resources, and industry) to military operations. The combination of strategic resources (the national sustainment base) and distribution processes (Service components) represents the Nation's total capabilities. These capabilities include the Department of Defense, the Military Services, other government agencies as necessary or appropriate, and the support of the private sector. Headquarters, Marine Corps and the Marine Corps supporting establishment plan and conduct Marine Corps strategic logistics—with the exception of aviation-peculiar support. Aviation-peculiar support is planned and conducted by the Chief of Naval Operations, the Navy supporting establishment, and the Naval Reserve. At the strategic level, the Marine Corps—

- Procures weapons and equipment, except aircraft and aviation ordnance.
- Recruits, trains, and assembles forces.
- Establishes bases, stations, and facilities to include ranges and airspace, to house, train, and maintain forces and stockpile resources.
- Mobilizes forces.
- Oversees and coordinates employment of strategic-level transportation assets.
- Regenerates forces.
- Provides command and control to manage the flow of resources from the strategic to the tactical level.

Operational Logistics

Operational logistics links tactical requirements to strategic capabilities to accomplish operational goals and objectives. It includes the support required to sustain campaigns and major operations. Operational logisticians assist in resolving tactical requirements and coordinating the allocation, apportionment, and distribution of resources within theater. They coordinate with logisticians at the tactical level to identify theater shortfalls and communicate these shortfalls back to the strategic source. The Marine Corps' operating forces, assisted by Headquarters, Marine Corps and the supporting establishment, are responsible for operational logistics. The Marine Corps component commander or the MAGTF commander in the absence of an in-theater Marine component commander, conducts operational logistics. The Marine Corps component commander may establish a MLC to perform operational logistic functions to support tactical logistics in the AO. The focus of operational logistics is to balance the MAGTF deployment, employment, and support requirements to maximize the overall effectiveness of the force. Marine Corps operational logistics orients on force closure, sustainment, reconstitution, and redeployment of Marine forces in theater.

Tactical Logistics

Tactical logistics includes organic unit capabilities and the combat service support activities that support military operations. It supports the commander's intent and concept of operations while maximizing the commander's flexibility and freedom of action. The goal of tactical level logistics is to prevent the MAGTF from reaching a premature culminating point. Tactical logistics involves the planning, execution, and coordination of the functions of logistics required to sustain and move units, personnel, equipment, and supplies. These functions must provide flexible and responsive combat service support to the MAGTF. Tactical logistics must be timely and anticipate the needs of the MAGTF to provide responsive support. Generally, the MAGTF and other Services and/or multinational forces assigned or attached to the Marine Corps component conduct tactical level logistic operations.

The MAGTF is task-organized with the combat service support organizations that it needs to accomplish assigned missions. Although no single MAGTF element has all of the operational and logistic capabilities to operate independently, each element has the capability for at least some basic self-support tasks. Marine tactical units have logistics officers and logistics sections, but the units that perform combat service support functions for these units are referred to as CSSEs. The CSSE provides general ground logistics to all MAGTF elements. The ACE possesses logistic capabilities essential for aircraft operations. Typically, the MAGTF deploys with accompanying supplies that enable it to conduct operations that range from 15 to 60 days, depending on the size of the MAGTF. Normally, resupply channels are established and flow of supplies initiated to support the MAGTF prior to the consumption of the MAGTF's organic supplies.

For more information on the levels of logistics see MCDP 4, *Logistics* and MCWP 4-1, *Logistics Operations*.

FUNCTIONS

Logistics is normally categorized in six functional areas: supply, maintenance, transportation, general engineering, health services, and services. Logistic systems and plans are usually developed to address each functional area. Logisticians commonly discuss support requirements and concepts in terms of these commodity areas. While each functional area is essential, all functions must be integrated into the overall logistic plan for total support of MAGTF operations.

Supply

Supply is the receipt, storing, issuing, and re-supplying of material necessary to conduct operations. The needs of the supported force drive all supply efforts. Supply systems should be organized to respond rapidly to changing tactical situations and flexible enough to accommodate modifications to the concept of operations.

Maintenance

Maintenance involves those actions taken to retain or restore materiel to serviceable condition. The purpose and function of equipment maintenance are universally applicable, but the Marine Corps has developed distinct applications for the support of ground-common and aviation-unique equipment.

Organizational-level maintenance is performed by the using unit on its organic equipment in ground and aviation units. The MAGTF combat service support units and organic maintenance organizations that possess intermediate maintenance capabilities conduct intermediate-level ground maintenance. Marine aviation logistics squadrons (MALS) conduct intermediate aviation maintenance. Depot-level maintenance for ground equipment, particularly Marine Corps-specific items, is performed at commercial and Department of Defense depot maintenance facilities. The Commander, Naval Air Systems Command, coordinates aviation, depot-level maintenance needs. Aviation maintenance support for a MEB may come from an intermediate maintenance activity or may be provided through a combination of MPS assets, fly-in support packages, and/or off-the-shelf spares or organic repair support from an aviation logistics support ship. While a MAGTF is aboard amphibious shipping, its aircraft maintenance support is provided by the ship's aircraft maintenance department, augmented by personnel from one or more of the MALS. Smaller MAGTFs draw support from MALS allowance lists (aviation consolidated allowance lists, consolidated allowance lists), fly-in support packages, and/or contingency support packages in a variety of combinations.

Transportation

Transportation is moving personnel and materiel from one location to another via highways, railroads, waterways, pipelines, oceans or air. For a MAGTF, transportation support is defined as that support needed to put personnel and materiel in the correct location at the proper time to start and maintain operations. The transportation system that supports an expeditionary MAGTF not only includes the means of transportation but also the methods to control and manage those transportation means.

General Engineering

General engineering supports the entire MAGTF. It involves a wide range of tasks that serve to sustain forward combat operations; e.g., vertical or horizontal construction and facilities maintenance. Most general engineering support for MAGTF ground units comes from the engineer support battalion of the force service support group. The combat engineer battalion provides combat and combat support engineering to the division. Engineering capabilities are also inherent in the Marine aircraft wing and are found in the Marine wing support squadrons of the Marine wing support groups. If MAGTF construction needs exceed its inherent engineering capabilities, the naval construction force may provide augmentation.

Health Services

The objective of health services is to minimize the effects of wounds, injuries, and disease on unit effectiveness, readiness, and morale. This objective is accomplished by a proactive, preventive medicine program and a phased health care system (levels of care) that extends from actions taken at the point of wounding, injury, or illness to evacuation to a medical treatment facility that provides more definitive treatment. Health service support deploys smaller, mobile, and capable elements to provide essential care in the theater. Health service support resources are flexible and adaptable and can be tailored to missions ranging from major theater wars to MOOTW.

Services

JP 4-0, *Doctrine for Logistic Support of Joint Operations*; Naval Doctrine Publication 4, *Naval Logistics*; and MCDP 4 discuss a variety of nonmateriel and support activities that are identified as services. These services are executed in varying degrees by each of the Military Services, the Marine Corps supporting establishment, and the MAGTF. An understanding of the division of labor and interrelationship of the responsibilities and staff cognizance for specific services is essential to ensure those services are provided. Typically, within the MEF, the force service support group provides the following services:

- Disbursing.
- Postal.
- Legal.
- Security support.
- Exchange.
- Civil affairs (augmentation required from the Marine Corps Reserves).
- Graves registration (United States Army provides technical expertise).

Centralization of these capabilities within the force service support group does not imply sole logistic staff cognizance for execution of the task. The Marine Corps component may absorb the legal services, civil affairs, and disbursing capabilities of the MAGTF. This allows the force service support group to facilitate host nation and coalition relations, contingency contracting, the development of inter-Service and multinational support agreements, as well as provide general support to the MAGTF.

JOINT AND MULTINATIONAL OPERATIONS

Marine forces often deploy for operations as part of a joint or multinational task force. Logistic support of joint or multinational operations may call for compliance with specific operational and administrative requirements that are unique to those operations. Current joint and multinational doctrine provides a standard frame of reference for the planning, direction, and conduct of joint or multinational operations.

Joint and multinational operations are complex and bring together diverse military organizations that must operate together and logistically support one force. Multinational forces may have differences in command and control systems, language, terminology, doctrine, and operating standards. The following considerations minimize the impact of this diversity and promote efficiency:

- Liaison is the basis for effective command and control of logistics in joint and multinational operations. Liaison representatives (e.g., liaison officers, liaison teams, couriers) are chosen specifically for their knowledge and familiarity with the capabilities, limitations, and logistic concept of operations of their Service/national organization.

- The demand for information often exceeds the capabilities of command and control equipment within joint and multinational commands. It is crucial that the commander identify, as early as possible, the command and control requirements that are external to the command or that require the use of national and/or host nation equipment.
- Standardization of logistic procedures by joint or multinational forces is essential.
- Agreements are made with probable joint or multinational partners regarding command and control of logistics. These agreements should cover principles, procedures, and overall logistic report requirements (including standard text format, standard databases, and data formats). Agreements should be arrived at by mutual agreement in advance of the operation.
- Joint or multinational forces adopt the procedures of one Service or nation if command and control agreements have not been determined in advance.
- The commander of all United States forces provides interpreters to facilitate command and control and ensure that United States interests are adequately protected.
- The operational acceptability and disclosure or release of communications security to allied governments for multinational operations will be determined and approved by national authorities before entering into discussions with allied nations.

In joint operations, the Services are normally responsible for providing their own logistics. However, the combatant commander, acting through the commanders of the component forces, is responsible for overall logistic coordination. The combatant commander must oversee the logistics for all parts of the unified force and may direct Marine Corps resources to support other Services. The combatant commander is specifically responsible for developing and sustaining military effectiveness by establishing an effective logistic structure. The combatant commander makes recommendations for joint efforts to improve economy consistent with military efficiency, reviews requirements, and recommends priorities and programs. He has the authority to coordinate the logistics of the Service components and to control distribution of that support when shortages occur. The most common type of support is single Service logistic support. However, plans may require or direct the use of other types of support such as common servicing, cross-servicing or joint servicing at the force, theater, military department, or Department of Defense-level.

The Marine Corps component commander may establish a MLC to provide or coordinate operational logistics as an interface between the MAGTF and joint logistic organizations and agencies. The MLC establishes the Marine Corps theater support structure to facilitate reception, staging, onward movement, and integration operations, as well as provide long-term logistic support for sustained operations ashore.

In multinational operations, logistics is a national responsibility. However, agreements may be made to establish the framework for mutual support. The exchange of logistic support between Unites States forces and alliance or coalition participants can create significant economies of effort and cost savings. This type of logistic support or cross-servicing may be in the form of supplies and/or services. Host-nation support agreements normally establish or specify the type and amount of such support. An acquisition cross-Service agreement is a nation-to-nation bilateral agreement that specifies procedures and points of contact for equal value exchange, replacement in kind or acquisition of goods and services between United States forces and the forces of the given nation.

If no appropriate international agreements exist, no authority exists whereby combatant commanders can provide for or accept logistic support from allies or coalition forces. Combatant commanders are not authorized to enter into logistic support relationships with other nations without direction from the National Command Authorities. Under these circumstances, and with National Command Authorities' approval, Unites States commanders should acquire as much logistical support as possible through tact and diplomacy, their knowledge of the other nation's doctrine, and personal and professional relations with the other nation's commanders and appropriate political leaders. In the absence of approved formal support agreements, authorization for Marine Corps forces to receive logistic support from the other nation's forces or to provide support to these forces must come from higher authority in the operational chain of command. This does not preclude normal contracting of goods and services from the local economy.

STRATEGIC LOGISTICS SUPPORT

Strategic Mobility

A central concept in the national military strategy is strategic mobility. Strategic mobility includes the ability of the United States armed forces to deploy expeditionary forces to any region in the world and sustain them for the full range of military operations. The strategic mobility triad is the combination of sealift, airlift, and pre-positioning. Figure 5-1 highlights the strengths of each component of the strategic lift triad.

	Closure Speed	Flexibility	Capability
Sealift		X	X
Pre-positioning	X		X
Airlift	X	X	

Figure 5-1. Strategic Mobility Considerations.

The sea dominates the surface of our globe. For any global power, sea power is essential. Lack of modern or developed infrastructure can pose significant problems for military action in the developing world. Many ports cannot handle the deepest-draft ships. Many airfields in the developing world cannot handle the largest military transport aircraft.

The expeditionary MEF is capable of rapid response as part of a naval amphibious force, MPF or ACF. In other words, the MAGTF moves to crisis areas via the strategic mobility triad: sealift, pre-positioning, and airlift.

United States Transportation Command

The Department of Defense single manager of the Defense Transportation System is the United States Transportation Command (USTRANSCOM). The Commander in Chief, USTRANSCOM provides air, land, and sea transportation and common-user port management at air/seaports of debarkation as well as air/seaports of embarkation for the Department of Defense across the range of military operations. USTRANSCOM is a unified command with the following component commands: Air Mobility Command, Military Traffic Management Command, and Military Sealift Command. The Commander in Chief, USTRANSCOM has command of the components and their assigned transportation assets in peace and war. The component commands organize, train, and equip their forces.

USTRANSCOM is the Department of Defense's single worldwide manager for common user ports of embarkation and debarkation. The single port manager concept ensures the seamless transfer of cargo and equipment in any given theater. Combatant commanders are the supported commanders in determining movement requirements and required delivery dates. USTRANSCOM is the supporting command that, with the transportation component commands, provides a complete movement system from origin to initial theater destination. This system includes use of military and commercial assets. It has the authority to procure commercial transportation services through component commands and to activate, with approval of the Secretary of Defense, the Civil Reserve Air Fleet, Ready Reserve Fleet, Sealift Readiness Program, and the Voluntary Intermodal Sealift Agreement. The USTRANSCOM component commands operate the Defense Transportation System. The specific operations of the Defense Transportation System are covered in JP 4-01, *Joint Doctrine for the Defense Transportation System*.

Department of Transportation

Under the National Plan for Emergency Preparedness (Executive Order 12656), the Secretary of Transportation leads the Federal transportation community. During national defense emergencies, the Secretary of Transportation has a wide range of delegated responsibilities, including executive management of the Nation's transportation resources in periods of crisis. A more detailed account of Department of Transportation responsibilities is contained in JP 4-01, chapter V.

Defense Logistics Agency

The Defense Logistics Agency, a strategic and operational level logistic agency of the Department of Defense, provides worldwide logistic support to the Military Departments and the combatant commands across the range of military operations, as well as to other Department of Defense components, federal agencies, foreign governments, or international organizations. It provides materiel and supplies to the Military Services and supports their acquisition of weapons and other equipment. Support begins with joint planning with the Services for parts for new weapon systems, extends through production and concludes with the disposal of materiel that is obsolete, worn out or no longer needed. The Defense Logistics Agency provides supply, support and technical and logistic services to all branches of the military.

OPERATIONAL LOGISTICS SUPPORT

Marine Corps Component

The Marine Corps component commander's logistic functions include the identification and coordination of required Marine Corps logistic support at the operational level. Assigned or attached Marine Corps forces forward their support requirements and priorities to the Marine Corps component commander. The Marine Corps component commander then determines what resources will be used to fulfill the requirements. The Marine Corps component commander develops agreements with other component commanders and participates in component, command-level working groups. In military operations other than war, logistic support may also apply to support of United States forces, other United States Government agencies, and forces of friendly countries or groups supported by United States forces.

During predeployment the Marine Corps component commander conducts force sustainment planning and force reception planning. Throughout deployment the Marine Corps component commander refines Marine Corps forces personnel, sustainment, transportation, and reception requirements. The Marine Corps component commander may meet these requirements using Service sources or other joint resources.

A key function of the Marine Corps component commander during employment is to inform the joint force commander of changes to personnel and logistic requirements that might affect the Marine Corps' ability to support the operation. During employment, the Marine Corps component commander concentrates on—

- Sustainment sourcing.
- Intratheater transportation asset allocation.
- Facility and base development.
- Host-nation support.
- Health services management.

During redeployment the Marine Corps component commander focuses on reconstituting Marine Corps forces. The identification of accurate mission costs and material losses is also important to the Marine Corps component commander.

Marine Logistics Command

The Marine Corps component commander may establish an MLC to fulfill his Service logistic responsibilities. The MLC is task-organized around a force service support group. When formed, it provides logistic support to all Marine Corps forces—and may provide limited support to other joint and multinational forces as directed by the combatant commander. The MLC provides operational logistics to Marine Corps forces as the Marine Corps component's logistics agency in theater. Operational-level logistics includes deployment, sustainment, resource prioritization and allocation, and requirements identification activities required to sustain the force in a campaign or major operation. These fundamental decisions concerning force deployment and sustainment are key for the MLC to provide successful logistical support.

The Marine Corps component commander provides the logistic policy for Marine Corps forces. The MLC executes that policy to support all Marine Corps forces. When priorities of support are required, the Marine Corps component commander provides these to the MLC commander. Likewise, the Marine Corps component commander ensures the MLC receives assistance and resources outside its organic capability; i.e., intelligence necessary for the MLC mission.

The Marine Corps component commander may employ the MLC when the following operational conditions occur:

- Expeditious force closure of a MEF-sized MAGTF is required.
- A MEF will be ashore for more than 60 days.
- Sequential maritime pre-positioning force offloads or backloads are planned or required.
- Common item or user support is planned.
- Theater logistic support is shallow or has shortfalls.

These conditions assist the Marine Corps component commander in deciding if a MLC is necessary and, if so, its composition and capabilities.

The MLC is task-organized to fit the mission and tailored to meet specific theater and situational requirements. The Marine Corps component commander establishes support relationships between the MLC and the MAGTF. The division of labor between the MLC and the MAGTF is theater-specific.

Reconstitution

Reconstitution is the responsibility of the Marine Corps component commander and is normally executed by the MLC. It is how commanders plan and implement the restoration of units to a desired level of combat effectiveness to meet mission requirements. While it transcends normal day-to-day force sustainment actions, it uses existing systems and units.

Commanders conduct reconstitution when units become combat ineffective or when shifting available resources can raise combat effectiveness closer to the level they desire. Besides normal support actions, reconstitution may include—

- Removing the unit from combat.
- Assessing it with external assets.
- Reestablishing the chain of command.
- Training the unit for future operations.
- Reestablishing unit cohesion.

Although reconstitution is largely a command and operations function, the actual refitting, supply, personnel fill, and medical actions are conducted by combat service support and administrative elements.

There are two methods for conducting reconstitution—reorganization and regeneration. Reorganization is action taken to shift internal resources within a degraded unit to increase its level of combat effectiveness. Regeneration is action taken to reconstitute a unit through significant replacement of personnel, equipment, and supplies in an attempt to restore a unit to full operational capability as rapidly as possible. Reorganization is normally done at the unit level and does not require extensive external support. Regeneration is normally accomplished by the Marine Corps component or MAGTF and involves augmentation from the supporting establishment.

When established, the MLC—augmented by the supporting establishment—is responsible for executing regeneration operations. The MLC will manage the inbound replacement equipment and combat replacement companies. It will provide storage for equipment, billeting for personnel, and coordinate movement of personnel and equipment to the major subordinate commands based on the Marine Corps component commander's priorities. The MLC coordinates the replacement of personnel using existing capabilities to move personnel. When an MLC is not established, the Marine Corps component commander will plan movement of replacement personnel and equipment from outside the theater directly to the appropriate units.

COMMAND AND CONTROL

All logistics efforts support mission accomplishment and are ultimately the responsibility of the commander. Command and control of logistics enables the commander to recognize requirements and provide the required resources. It must provide visibility of both capabilities and requirements. This visibility allows the commander to make decisions regarding the effective allocation of scarce, high-demand resources. Additionally, command and control facilitates the integration of logistic operations with other warfighting functions so that the commander's time for planning, decision, execution, and assessment is optimized. Only when command and control effectively direct the logistic effort can logistics effectively and efficiently support the mission.

At the strategic level of war, the Marine Corps must disseminate information and directives to and from Headquarters, Marine Corps. The principal agents for dissemination of logistic policy and information are Deputy Commandant for Installations and Logistics and Deputy Commandant for Aviation. The

Commander, Marine Corps Materiel Command is responsible for executing Marine Corps strategic-level ground logistics, while Commander, Naval Air Systems Command executes strategic-level aviation support. An effective flow of information and directives enables the Marine Corps to manage materiel readiness, mobilization and deployment support, and materiel replenishment.

At the operational level of war, the logistic effort enables force closure; establishes and maintains arrival and assembly areas; and coordinates intra-theater airlift, sustainment needs, and force reconstitution and redeployment requirements. Marine Corps component commanders direct logistics at the operational level.

Marine Corps component G-4s coordinate ground logistics issues with subordinate MAGTF G-4/S-4s. Commander, Naval Air Force, Pacific, and Commander, Naval Air Force, Atlantic, deal directly with the MAGTF ACE aviation logistics department to assist in resolving aviation logistic requirements. In joint operations, the principal logistics agent is the J-4 at the unified, sub-unified, and/or the joint task force staff level.

LOGISTICS SUPPORT TO MAGTF OPERATIONS

Successful deployment, sustainment, employment, and redeployment of a MAGTF are the result of well-coordinated logistic support activities conducted at the strategic, operational, and tactical levels. The organization of forces, materiel support, and assigned logistic responsibilities are structured with one goal—to logistically support MAGTF operations. Initially, logistics is drawn from internal Marine Corps/Navy resources located within the operating forces, the Marine Corps Reserve, and the supporting establishment. Specific operational requirements dictate the extent to which additional logistic support is drawn from other Services, non-Department of Defense resources, and multinational resources.

JP 1-02 defines combat service support as the essential capabilities, functions, activities, and tasks necessary to sustain all elements of operating forces in theater at all levels of war. Within the national and theater logistic systems, it includes but is not limited to that support rendered by service forces in ensuring the aspects of supply, maintenance, transportation, health services, and other services required by aviation and ground combat troops to permit those units to accomplish their missions in combat. Combat service support encompasses those activities at all levels of war that produce sustainment to all operating forces on the battlefield. Combat service support in the Marine Corps is a function or tasking associated with a unit that, by table of organization and table of equipment, is organized, equipped, and trained as a combat service support organization to perform those operations.

The Marine Corps' logistic mission is to generate MAGTFs that are rapidly deployable, self-reliant, self-sustaining, and flexible and that can rapidly reconstitute. The Marine Corps, in coordination and cooperation with the Navy, has made logistical self-sufficiency an essential element of MAGTF expeditionary warfighting capabilities.

Marine Expeditionary Force

The MEF is normally supported by at least one force service support group as the CSSE. It is designed to deploy with accompanying supplies for up to 60 days. The force service support group is task-organized to provide a full range of support functions from sea bases aboard naval shipping or from expeditionary bases ashore. It can task-organize the appropriate capability into smaller combat service support groups/detachments in direct support of specific task-organized units of the MEF. The force service support group is the embodiment of a fundamental principle—economy of operations through centralization of logistic resources and decentralization in executing support operations.

The force service support group, as the CSSE of a MEF, is organized to provide a full range of combat service support capabilities for 60 days. Normally it is comprised of the following battalions:

- Headquarters and service battalion.
- Maintenance battalion.
- Engineer support battalion.
- Supply battalion.
- Transportation support battalion.
- Medical battalion.
- Dental battalion.

The Marine division depends on the force service support group as its primary source of logistic support. However, the division has organic combat engineer capability that it can employ in general and direct support of division organizations before requesting additional support from the engineer assets of the force service support group. The division also possesses a limited general and direct support motor transport capability, which is normally employed in support of the division headquarters.

The Marine aircraft wing possesses logistic units in the Marine wing support groups that provide aviation ground support. The Marine Aviation Logistics Squadron (MALS) provides intermediate-level, aircraft-specific aviation supply, maintenance, avionics, and ordnance capabilities in direct support of aircraft squadrons and groups. The Marine aircraft wing depends on the force service support group as its primary external source of ground logistics and for delivery of aviation bulk commodities.

Marine Expeditionary Brigade

The MEB, with up to 30 days of supplies for sustained operations, can operate independently or serve as the advance echelon of a MEF. The brigade service support group is task-organized to provide combat service support to the MEB. It is structured from personnel and equipment of the force service support group. Normally the brigade service support group is comprised of the following units and may establish CSS detachments as required.

- Headquarters company.
- Transport support company.
- Supply company.
- Maintenance company.
- Engineer support company.
- Bulk fuel company.
- Collecting and clearing company.
- Surgical support company.
- Dental detachment.

Marine Expeditionary Unit (Special Operations Capable)

The MEU(SOC) is the standard forward-deployed Marine expeditionary organization. The MEU service support group is task-organized to provide combat service support to the MEU. The MEU service support group provides 15 days of sustainment, except for aviation. 90 days of supplies and repair parts are normally provided aboard amphibious shipping, except for aviation ordnance, which is constrained to 15 days due to limits of amphibious shipping explosive material storage capabilities.

Special Purpose MAGTF

A special purpose MAGTF is organized to accomplish a specific mission, operation, or exercise. The size and capabilities of its elements will vary with the mission. Normally, a combat service support detachment is task-organized to be the CSSE. The assigned mission will reflect the logistic functional capabilities and the level of organic sustainment assigned.

Air Contingency Force

An ACF consists of air-deployable forces that are maintained in Marine Corps Forces, Pacific, and Marine Corps Forces, Atlantic. The size of the force can range from a reinforced rifle company to a regimental-sized force with an appropriately sized CSSE.

Maritime Pre-positioning Forces

MPFs provide an added dimension to strategic mobility, readiness, and global responsiveness. By pre-positioning the bulk of equipment and 30 days of supplies for a notional 17,600-man force aboard specially designed ships, the MPF program reduces MAGTF response time from weeks to days. Included in each MPS squadron is organizational-level, common aviation support equipment and limited, intermediate-level support equipment. The ACE deploys with a fly-in support package that, when combined with pre-positioned assets on the MPS squadron, provides critical aviation support for 30 days of combat flying. Equipment and supplies, to a limited extent, can be offloaded selectively to support smaller MAGTFs. While normally associated with MEB employment, supplies and equipment aboard MPS squadrons are loaded to support lesser contingencies if necessary.

Aviation Logistics Support Ship

The aviation logistics support ship is a program developed to transport critical, tailored, intermediate-level maintenance and supply support to a forward operating area in support of deployed aircraft. The two aviation logistics support ships (one located on the west coast and one on the east coast) are under the ADCON of the Military Sealift Command. They provide a dedicated sealift for movement of the Marine aviation logistics squadron supplies and equipment and an afloat intermediate maintenance activity capability. This intermediate maintenance activity is task-organized to repair aircraft parts and equipment of the aircraft platforms within the MAGTF.

Norway Geopre-positioning Program

The Norway Geopre-positioning Program is a capability similar in scope to that of an MPS squadron. The program, established with the Government of Norway, permits the pre-positioning and maintenance of a MEB's worth of equipment in underground storage facilities in Norway.

War Reserve Materiel Support

Availability and responsiveness of access to stocks of war reserve materiel is maintained in accordance with Department of Defense policy. War reserve materiel is defined as mission-essential principal end items, secondary items, and munitions required to attain operational objectives in the scenarios authorized for sustainability planning and other stockage objectives approved for programming in the Defense Planning Guidance. War reserve materiel inventories are acquired during peacetime. These inventories are flexible. They provide an expansion capability that can respond to a spectrum of regional contingencies while minimizing investment in resources. Authority to approve the release of war reserve materiel stocks is limited to specified commanders.

Tactical Level Command and Control

Command and control of logistics at the tactical level focuses on monitoring, directing, and executing logistic operations and maintaining communications with supporting operational level forces. MAGTF element G-4s/S-4s employ all of their organic logistic support capabilities and coordinate with the CSSE for additional support as necessary. The CSSE's mission is to provide logistics exceeding the organic capabilities of other MAGTF elements. The ACE aviation logistics department works with the MALS to resolve aviation logistics shortfalls.

SUPPORTING ESTABLISHMENT

The organization of the Marine Corps consists of Headquarters, Marine Corps; the operating forces; the Marine Corps Reserve; and the supporting establishment. Each category has inherent logistic capabilities and specific logistic responsibilities at the strategic, operational, and tactical levels of war. The supporting establishment, referred to as the "fifth element" of the MAGTF, is inextricably linked to the operational forces as it furnishes logistics vital to the overall combat readiness and sustainment of the Marine Corps. Its inherent logistics capabilities and specific logistics responsibilities are found throughout the strategic and operational levels of war. It is the source of Marine Corps strategic logistics in that it recruits, trains, equips, and provisions the force, as well as providing the bases and stations with the associated infrastructure to house, train, and maintain the force. The supporting establishment provides the operational logistics link between the strategic and tactical levels. Although not part of the Marine Corps, the Navy supporting establishment also provides essential logistics to the Marine Corps; in particular its support of Marine aviation.

Marine Corps Combat Development Command

Logistics is a significant focus of the Marine Corps Combat Development Command (MCCDC). Each division within MCCDC establishes logistics branches and sections to ensure that logistics is properly integrated into the Concepts Based Requirements Process, the Combat Development System, and the Marine Corps Capabilities Plan. MCCDC develops or identifies the concepts, doctrine, organizations, equipment requirements, training programs, facilities, and support that generate Marine Corps' warfighting capabilities. It is also responsible for professional military education programs that teach Marines their warfighting profession and provide an intellectual environment for improving established methods and equipment for mission accomplishment.

Marine Corps Materiel Command

The Commander, Marine Corps Materiel Command is the single process owner for the Marine Corps life cycle management process. He executes this process through two subordinate commanders: Commander, MARCORSYSCOM and Commander, Marine Corps Logistics Bases.

MARCORSYSCOM is responsible for research, development, acquisition, and life-cycle management of Marine Corps funded materiel and information systems. MARCORSYSCOM directs Marine Corps-sponsored programs and represents the Marine Corps in the development of other Service-sponsored programs in which the Marine Corps participates. MARCORSYSCOM coordinates program interface internally within the Marine Corps and externally with Department of the Navy, Department of Defense, other Services, Congress, and industry. MARCORSYSCOM also manages Marine Corps ground ammunition acquisition programs and Marine Corps owned and controlled ground ammunition stocks. The ground ammunition function is particularly significant in supporting MAGTF sustainability during operations and crisis action response planning and execution.

The Marine Corps logistics bases at Albany, Georgia, and Barstow, California, and the Blount Island Command at Jacksonville, Florida, are under the command of the Commander, Marine Corps Logistics Bases, headquartered at Marine Corps Logistics Base, Albany. They provide general, Service-level supply and maintenance support to the Marine Corps and certain support services to the Department of Defense. Marine Corps Logistics Base, Albany, is the inventory control point for the Marine Corps supply system. Albany and Barstow have materiel storage facilities that house consumable and repairable materiel, including some pre-positioned war reserve materiel. They control the storage of principal end items. They also have repair centers that perform depot-level maintenance and, when directed, overflow intermediate level maintenance on ground equipment. They can also serve as manufacturing centers when directed. Blount Island Command is responsible for inventory management and equipment maintenance, modification, and replacement in support of the MPF and the Norway Geopre-positioning Program.

Marine Corps Bases, Stations, and Reserve Support Centers

Marine Corps bases, stations, and reserve support centers provide the infrastructure and facilities that support the operating forces and allow training as a MAGTF. They provide the programs that ensure safety and services for all Marines, their families, civilian employees, and retirees. The bases, stations, and centers provide the training areas, airspace, and infrastructure that enable the operating forces, including the Marine Corps Reserve, to maintain their combat readiness. They support force deployments on routine and contingency response operations and provide critical logistics to deploying units from predeployment preparation through deployment and reconstitution. Finally, in support of the total force, some bases and stations are designated stations of initial assignment for Marine Corps Reserve mobilization and are responsible for assisting the operating forces with the throughput of Marine Corps Reserve personnel and materiel in support of MAGTF deployment.

Department of the Navy Agencies

Certain Department of the Navy agencies support both the Navy and the Marine Corps. In logistics, the most visible functions are naval aviation materiel support and health service support. Materiel support is provided by Naval Systems Command. The Navy's logistics effort is divided between several commands. Naval Supply Systems Command provides logistics support less aviation logistics support. Naval Sea Systems Command oversees the five depot maintenance shipyards. Naval Air Systems Command provides aviation logistics support to include oversight of the three aviation maintenance depots. Other naval organizations providing significant supply support include Naval Medical Logistics Command, which is the Navy and Marine Corps subject matter expert for medical materiel, and procures all medical and dental equipment, services, and supplies for naval forces. Naval Facilities Engineering Command provides initial outfitting of chemical, biological, and radiological defense material and equipment to overseas shore installations, Naval Construction Force, and naval beach group units. Space and Naval Warfare Systems Command, Chesapeake provides software support for the fleet logistics programs that automate supply, inventory control, maintenance, and financial management.

FUTURE SUSTAINMENT OPERATIONS

MAGTFs conducting expeditionary operations require sustainment that is responsive, flexible, and scalable to generate and sustain combat power. Key to sustainment operations in the future will be the use of sea-based logistics that provide a secure base for the rapid and anticipatory distribution of supplies and services. Enhancements in information technology will be used to ensure visibility of the location and availability of supply items, exercise economy, and streamline the prioritization and delivery of services to support rapidly moving maneuver and fire support units.

Sea-basing of sustainment will allow the commander to further reduce the footprint of sustainment personnel and material ashore, reducing the need for security forces, and minimizing the impact of MAGTF operations on the local population in the AO. Precision logistics will ensure that only the material and services actually needed by the forces will be brought ashore. Marine Corps forces will be able to rapidly deploy to the AO, conduct operations, and redeploy, ready for subsequent operations.

CHAPTER 6

Planning and Conducting Expeditionary Operations

Contents	
Maneuver Warfare	6-2
Operational Design	6-3
Visualize	6-5
Describe	6-5
Direct	6-6
Planning	6-9
Commander's Battlespace Area Evaluation	6-10
Analyze and Determine the Battlespace	6-10
Centers of Gravity and Critical Vulnerabilities	6-11
Commander's Intent	6-12
Commander's Critical Information Requirements	6-13
Commander's Guidance	6-14
Mission	6-15
Decisive Action	6-17
Shaping Actions	6-18
Battlefield Framework	6-19
Integrated Planning	6-19
Single Battle	6-20
Deep Operations	6-21
Close Operations	6-22
Rear Operations	6-22
Noncontiguous and Contiguous	6-23
Main and Supporting Efforts	6-24
The Reserve	6-27
Security	6-28
Phasing	6-29
Operation Plans and Orders	6-30
Transitioning Between Planning and Execution	6-31

Execution	6-31
Command and Control	6-32
Assessment	6-33
Tactical Tenets	**6-34**
Achieving a Decision	6-35
Gaining Advantage	6-36
Tempo	6-38
Adapting	6-39
Exploiting Success and Finishing	6-40

"The art of war is simple enough. Find out where your enemy is. Get at him as soon as you can. Strike at him as hard as you can, and keep moving on."
—Lieutenant General U.S. Grant, USA

"To move swiftly, strike vigorously, and secure all the fruits of victory is the secret of successful war."
—Lieutenant General Thomas "Stonewall" Jackson, CSA

This chapter provides the foundation for MAGTF tactical operations because it discusses the importance of single battle, decisive and shaping actions, centers of gravity and critical vulnerabilities, main and supporting efforts, security, and the reserve. It describes operational design and addresses how the MAGTF commander plans and conducts expeditionary operations. It also identifies the tactical tenets essential to succeeding on the battlefield.

MANEUVER WARFARE

The Marine Corps practices maneuver warfare when conducting operations. MCDP 1 defines maneuver warfare as a warfighting philosophy that seeks to shatter the enemy's cohesion through a variety of rapid, focused, and unexpected actions which create a turbulent and rapidly deteriorating situation with which the enemy cannot cope. Maneuver warfare is based on the avoidance of the enemy's strengths—surfaces—and the exploitation of the enemy's weaknesses—gaps. Rather than attacking the enemy's surfaces, Marines bypass the enemy's defense and penetrate those defenses through gaps to destroy the enemy system from within. The goal of maneuver warfare is to render the enemy incapable of effective resistance by shattering his moral, mental, and physical cohesion.

Maneuver provides a means to gain an advantage over the enemy. Traditionally, maneuver has meant moving in a way that gains positional—or spatial—advantage. For example, a force may maneuver to envelop an exposed enemy flank or deny him terrain critical to his goals. The commander may maneuver to threaten the enemy's lines of communications and force him to withdraw. He may maneuver to seize a position that brings effective fire to bear against the enemy but protects his forces against enemy fires.

To maximize the usefulness of maneuver, the commander must maneuver his forces in other dimensions as well. The essence of maneuver is taking action to generate and exploit some kind of advantage over the enemy as a means of accomplishing his objectives as effectively as possible. That advantage may be psychological, technological or temporal as well as spatial.

A force maneuvers in time by increasing relative speed and operating at a faster tempo than the enemy. Normally, forces maneuver both in time and space to gain advantage and, ultimately, victory at the least possible cost. Operation Desert Storm (1991) is a classic example where forces used time and space to their advantage to outmaneuver the enemy. While the Marines and Arab coalition forces fixed the Iraqis in Kuwait, other coalition forces rapidly maneuvered through the Iraqi desert to out flank the enemy. The success of this maneuver led to the complete disruption of the Iraqi defense and their rapid capitulation. This operation illustrates the importance of synchronizing the maneuver of all MAGTF elements to achieve a decision.

OPERATIONAL DESIGN

Commanders initiate the conduct of operations with a design that will guide their subordinate commanders and the staff in planning, execution, and assessment. This operational design is the commander's tool for translating the operational requirements of his superiors into the tactical guidance needed by his subordinate commanders and his staff. The commander uses his operational design to visualize, describe, and direct those actions necessary to achieve his desired end state and accomplish his assigned mission. It includes the purpose of the operation, what the commander wants to accomplish, the desired effects on the enemy, and how he envisions achieving a decision. Visualization of the battlespace and the intended actions of both the enemy and the friendly force is a continuous process that requires the commander to understand the current situation, broadly define his desired future situation, and determine the necessary actions to bring about the desired end state. The commander then articulates this visualization to his subordinate commanders and staff through his commander's battlespace area evaluation (CBAE) and guidance. By describing his visualization in this concise and compelling method, the commander focuses the planning and execution of his subordinate commanders and staffs. Finally, the commander directs the conduct of operations by issuing orders, assigning missions and priorities, making decisions, and adjusting his planned actions as necessary based on assessment. See figure 6-1 on page 6-4.

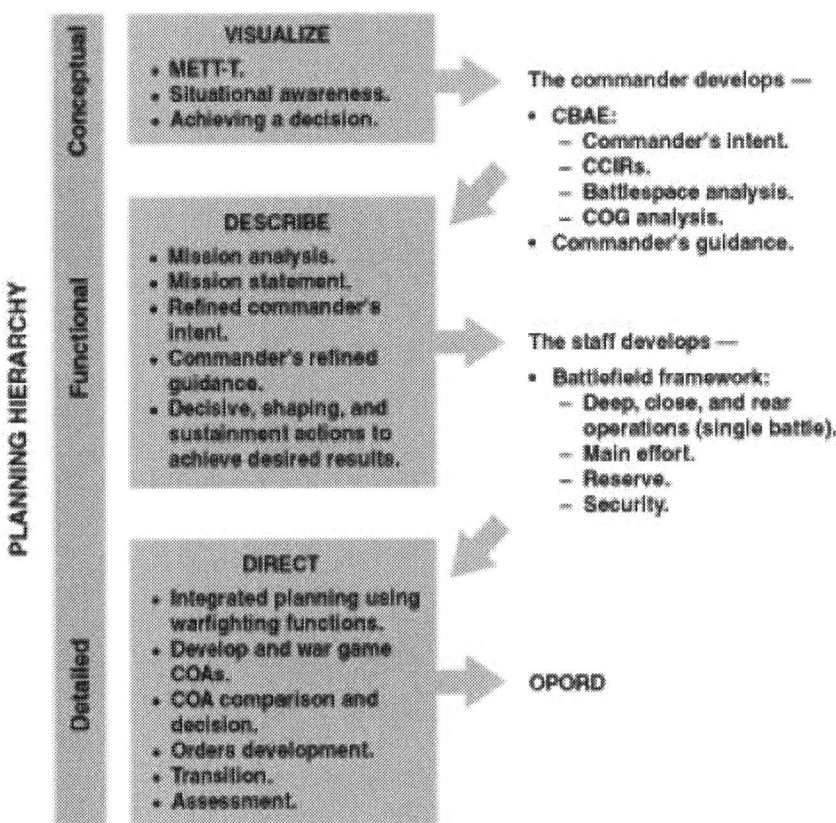

Figure 6-1. Operational Design.

Operational design differs at various levels of command, principally in the scope and scale of operations. Higher level commanders, such as the component and MAGTF commander, identify the time, space, resources available, and purpose of operations that support the joint force commander's campaign plan or component commander's operational design. At a lower level of command, the commander may be able to include in his operational design a detailed description of the battlespace, objectives, available forces and desired task organization, and guidance on the phasing of the operation. The elements of operational design include—

- Factors of METT-T.
- CBAE consisting of the commander's analysis of the battlespace, commander's intent, center of gravity analysis, and commander's critical information requirements.
- Commander's guidance, to include desired effects.
- Decisive actions.
- Shaping actions.

- Sustainment.
- Principles of war and tactical fundamentals. See appendix B.
- Battlefield framework.
- Operation plan or order.

Visualize

The visualize portion is what MCDP 5, *Planning*, refers to as conceptual planning, the highest level of planning. The commander determines the aims and objectives of the operation. During visualization, the first task for the commander is to understand the situation. He studies the situation to develop a clear picture of what is happening, how it got that way, and how it might further develop. The commander considers the information available on the factors of METT-T and any other information on the situation or potential taskings from higher headquarters. He develops an initial view of friendly actions, desired effects and their results, and determines the means to achieve those results. Part of the commander's thinking should also include assuming the role of the enemy, considering what the enemy's best course of action may be, and deciding how to defeat it. Thinking through these factors helps the commander develop increased situational awareness. The commander must also address possible outcomes and the new situations that will result from those possibilities. As the situation changes, so will the solution and the actions that derive from it. Combining this initial understanding of the situation within the battlespace with his experience and military judgment, he may begin his visualization by posing the following questions:

- Where am I? Where is the enemy?
- Where are my friends? Where are the enemy's friends?
- What are my strengths? What are the enemy's strengths?
- What must I protect? What are the enemy's weaknesses?
- What must I do and why? What will the enemy do and why?
- What is the enemy's most dangerous course of action?

As the commander considers these questions, he visualizes what he thinks he has to accomplish to achieve a decision and best support his higher commander's operation. This becomes the basis for his CBAE and guidance he provides to his subordinate commanders and the planners in the describe portion of operational design.

Describe

The describe portion is a combination of conceptual planning and what MCDP 5 refers to as functional planning, the middle level of planning where the commander and the staff consider discrete functional activities that form the basis for all subsequent planning. It begins when the commander articulates his vision through his CBAE and initial guidance. The commander then uses this visualization to focus and guide the staff as they conduct mission analysis to determine the mission of the force. Mission analysis provides the commander and his staff with additional insight on the situation. Combined with any intelligence or operational updates, mission analysis may prompt the commander to refine his vision, confirming or modifying his commander's intent or other initial guidance on decisive and shaping actions and sustainment. See figure 6-2.

Once the mission statement has been produced, the commander and staff are ready to further develop the operational design by describing how the command will achieve a decision through decisive and shaping actions that accomplish the mission and achieve the desired effects. They also describe how these actions will be sustained. Receiving necessary commander's planning guidance, the staff begins to develop the battlefield framework. See figure 6-3 on page 6-8. It describes how the commander will organize his battlespace and his forces to achieve a decision. The battlefield framework consists of the battlespace organization of envisioned deep, close, and rear tactical operations as well as the organization of the force into the main effort, reserve, and security. Supporting efforts are addressed in the context of deep, close, and rear operations as part of the single battle.

Direct

The direct portion is a combination of functional planning and what MCDP 5 refers to as detailed planning, the lowest level of planning. During direct, the commander and the staff determine the specifics of implementing the operational design through the operation plan or order. Armed with the description of how the commander intends to achieve a decision and obtain his desired end state, planners conduct integrated planning using the battlefield framework and the six warfighting functions to develop and war game courses of action that address the following considerations and issues:

- Type of operation.
- Forms of maneuver.
- Phasing/sequencing of the operation.
- Security operations.
- Sustaining the operation.
- IO.
- Targeting priorities.
- Intelligence collection priorities.
- Assessment.

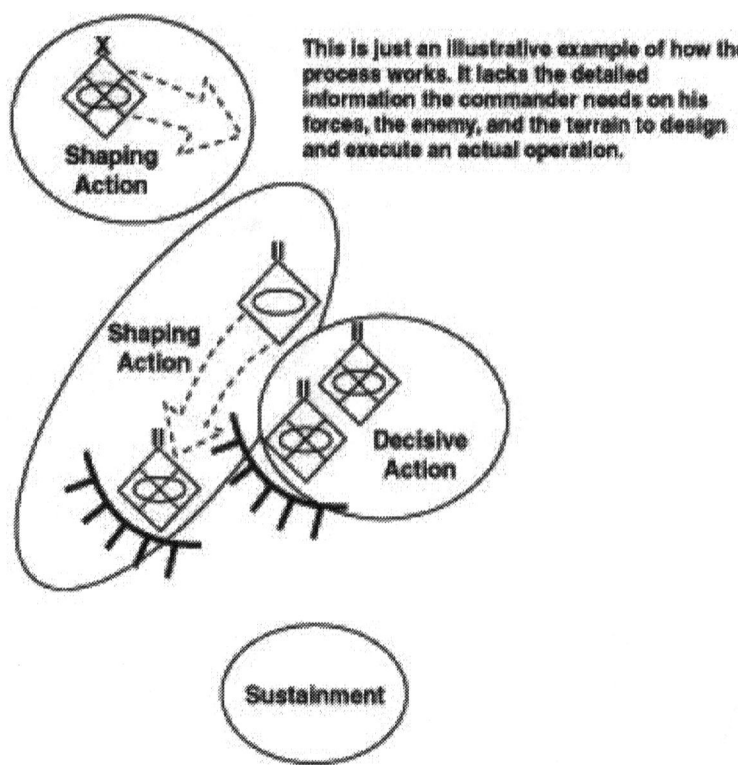

Figure 6-2. Commander's Vision of Decisive
and Shaping Actions and Sustainment.

The purpose of this operation is to defeat the enemy's first tactical echelon. I see the enemy's tactical strength as his mobile reserves. I cannot let the enemy commit these reserves in a decisive manner. To support the higher commander's plan, I will have to keep the reserve mechanized brigade from committing against our higher commander's main effort or being used decisively against my forces. I want to shape the enemy by having him first commit his reserve armor battalion against my secondary effort. Simultaneously, by using lethal and nonlethal fires, I want to control the timeline for the commitment of the enemy's reserve mechanized brigade and once committed against my forces, I want to limit its capability. These shaping actions will allow me to fix the enemy reserves while I mass my combat power at the time and place of my choosing. I want to exploit my tactical center of gravity—my superior tactical mobility—and combined arms. I want to avoid the enemy's fixed defenses and focus my decisive action against the enemy's flank to defeat the two isolated mechanized battalions. Once defeated, I want to rapidly focus on the defeat of his remaining mechanized and reserve units that were fixed by my supporting effort. I want a viable security force protecting the flank of my main effort. My sustainment must be task-organized and positioned forward to allow the force to maintain operational momentum.

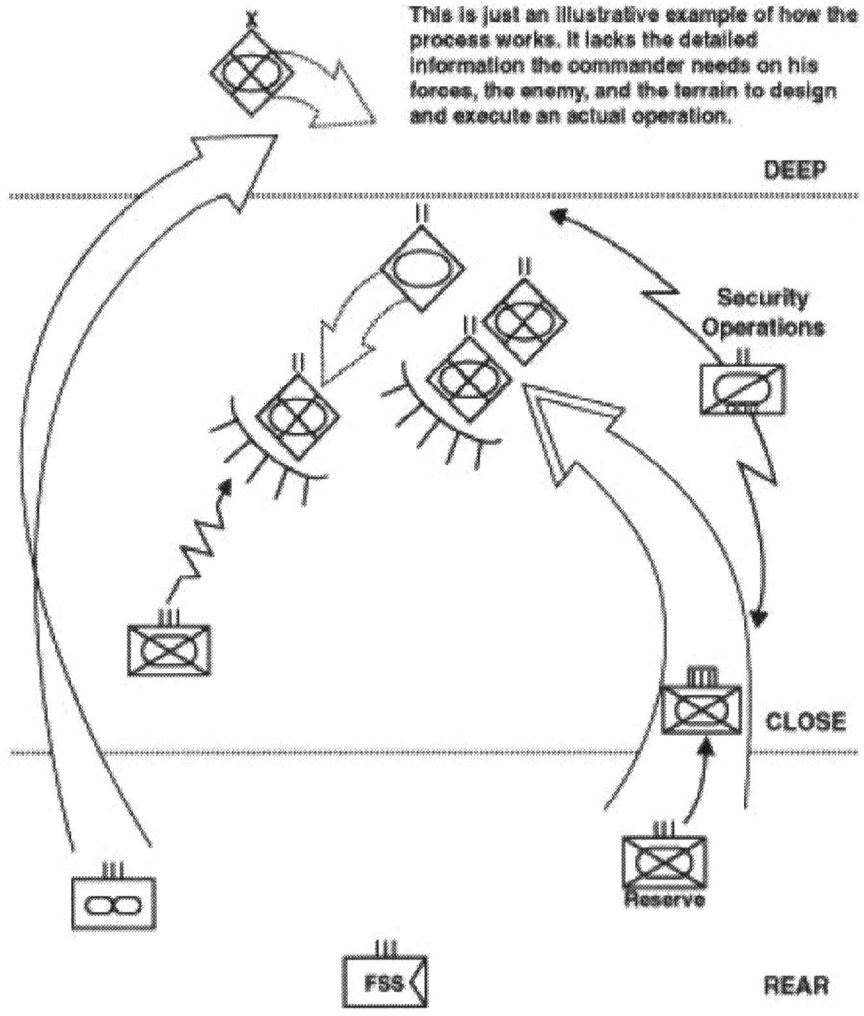

Figure 6-3. Battlefield Framework.

As this integrated planning continues, the commander chooses a course of action and, if time and situation allow, the staff conducts detailed planning to provide further direction to the force and prepare necessary operation plans and orders. Once the plan or order is completed, the direct portion of operational design concludes with the transition of the plan or order to the subordinate commanders and the staff that will execute it. The operational design, once developed into an operation plan or order, is the basis for execution and aids the commander and the staff as they execute operations.

The commander assesses the success of the operation by comparing the envisioned operational design—as expressed in the operation order—with what is actually occurring in the battlespace. If the assessment indicates the need to modify or adjust the operational design, the commander will again visualize what must be done and then he and the staff will describe how it will be accomplished by modifying or adjusting the battlefield framework. Fragmentary orders, branch plans or sequels to direct the operation will be prepared and issued, if necessary.

PLANNING

As described in MCDP 5, planning is the art and science of envisioning a desired future and laying out effective ways of bringing it about. It encompasses envisioning this desired end state and arranging a configuration of potential actions in time and space that will realize the end state. Planning is an essential element of command and control and the responsibility to plan is inherent in command. It is a truism that planning is half of command and control. The fundamental object of command and control is also the fundamental object of planning—to recognize what needs to be done in any situation and to ensure appropriate actions are taken. This requires commanders who can visualize what they want to happen, the effects they want to achieve, and how they will employ their forces to achieve their goals. They must be able to contemplate and evaluate potential decisions and courses of action in advance of taking action. The commander must be constantly aware of how much time a situation allows for planning and make the most of that available time. Whether planning is done deliberately or rapidly, the commander must display an acute awareness of the time available. All planning is time sensitive.

Planning can be viewed as a hierarchical continuum with three levels of planning—conceptual, functional, and detailed. All three levels are used at various times in the Marine Corps Planning Process (MCPP), such as the use of conceptual planning during mission analysis, functional planning during course of action development, and detailed planning during orders development.

While the commander is primarily involved at the conceptual level, he is an important participant throughout the planning process and must supervise his subordinate commanders and the staff in their efforts at the functional and detailed levels of planning. The commander's conceptual planning provides the basis for all planning performed by the staff. He must organize and train his staff to gather, manage, and process information essential to the commander's decisionmaking process. The size and capabilities of the staff depend on the level of command and the information and decisionmaking needs of the commander.

The MCPP is the vehicle the Marine Corps component or MAGTF commander and their staffs use to provide input to the joint planning process. It interfaces with the joint planning system during the supporting plan development phase in deliberate planning and during the situation development phase during joint crisis action planning. The Army's planning process closely resembles the MCPP, enabling close coordination in planning and execution among Marine Corps and Army forces assigned supporting missions or attached to the other Service. MAGTF planners must ensure that IO planning begins at the earliest stage of operation planning, is nested within the IO plans of the higher headquarters, and fully integrated into the MAGTF operations plan.

COMMANDER'S BATTLESPACE AREA EVALUATION

CBAE is the commander's personal vision based on his understanding of the mission, the battlespace, and the enemy. It is his visualization of what needs to be done and his first impressions of how he will go about doing it. He uses CBAE to articulate his initial view of the operational design. This visualization is used to transmit critical information to subordinate commanders and the staff and is the basis for the commander's planning and decisionmaking. It identifies the commander's battlespace, center of gravity and critical vulnerabilities, the commander's intent, and his critical information requirements. The staff normally assists the commander in preparing much of his CBAE, including battlespace appreciation, center of gravity analysis, and determining commander's critical information requirements. The G/S-2 is particularly helpful to the commander in determining possible enemy centers of gravity.

Analyze and Determine the Battlespace

As described in chapter 4, the commander's battlespace consists of his AO, area of influence, and area of interest. The commander analyzes his assigned AO, comparing the capabilities of his forces with the mission assigned to determine his area of influence. He then visualizes how he will use his forces within the battlespace to accomplish his mission. This visualization allows the commander to recognize critical information requirements that will determine the extent of his area of interest.

The commander compares his AO to his area of influence to determine whether the AO's size and location will allow him to accomplish his mission. If the AO is too small to allow the commander to use all the assets of the MAGTF effectively to accomplish the mission, then he should request a larger AO be assigned. If he determines that the AO assigned is too large for his force or that it is not located to best accommodate the MAGTF, then he should request a new or modified AO, or additional forces, from his commander. Regardless of its size, the MAGTF commander must be able to command and control his forces throughout the assigned AO.

Centers of Gravity and Critical Vulnerabilities

The commander continues to visualize what he must do and how he thinks he will use his force to accomplish that mission. An important aspect of the commander's visualization includes his analysis of centers of gravity and critical vulnerabilities. This analysis, based on the expected enemy COA, assists the commander in visualizing the relative strengths and weaknesses of the enemy and friendly forces.

As discussed in MCDP 1, a center of gravity is an important source of strength. Both enemy and friendly forces have centers of gravity. Depending on the situation, centers of gravity may be intangible characteristics such as resolve or morale. They may be capabilities such as armored forces or aviation strength. They may be the cooperation between two arms, the relations in an alliance or a force occupying key terrain anchoring an entire defensive system. Employing friendly strengths or centers of gravity to directly attack the enemy's strength should be avoided whenever possible. Rather, the commander seeks to employ his strength against threat weaknesses. To accomplish this task, the commander must identify the enemy's critical vulnerabilities; i.e., vulnerabilities that permit destruction of a capability without which the enemy cannot function effectively. Attacking critical vulnerabilities may achieve effects that bend the enemy to the commander's will.

Critical vulnerabilities provide an aiming point for the application of friendly strengths against threat weaknesses. The commander directs his force's strength at those capabilities that are critical to the enemy's ability to function—to defend, attack or sustain himself or to command his forces. The commander focuses on those critical vulnerabilities that will bend the enemy to his will most quickly. He must establish a process to identify those capabilities that are vulnerable and whose destruction or disruption will achieve the desired results. Once identified, critical vulnerabilities assist the commander in choosing where, when, and what will constitute decisive action. By attacking critical vulnerabilities, the commander increases the potential that the attack may in fact be the decisive action. Friendly critical vulnerabilities must also be identified to protect the friendly center of gravity from similar attack by the enemy.

The commander's analysis of centers of gravity and critical vulnerabilities during CBAE may require refinement as more information about the enemy and the tactical situation becomes available. The commander will continue to refine his visualization of the battlespace and his mission, which may require him to modify or delete his current choice for centers of gravity and critical vulnerabilities. Center of gravity and critical vulnerability analysis is an ongoing process and the commander's thinking on these items during CBAE may be radically altered during the remainder of the planning process and once the plan is executed.

Commander's Intent

The commander continues his CBAE by describing the interaction of the enemy, his own force and the battlespace over time and how he will achieve a decision that leads to the desired end state. He communicates this vision to his subordinates through the most important element of CBAE—commander's intent.

As described in MCDP 1, commander's intent is the commander's personal expression of the purpose of the operation. It must be clear, concise, and easily understood. It may also include how the commander envisions achieving a decision as well as the end state, conditions, or effects that, when satisfied or achieved, accomplish the purpose.

Commanders intent helps subordinates understand the larger context of their actions and guides them in the absence of orders. It allows subordinates to exercise judgment and initiative—in a way that is consistent with the higher commander's aims—when the unforeseen occurs. This freedom of action, within the broad guidance of the commander's intent, creates tempo during planning and execution.

Higher and subordinate commander's intent must be aligned. Commander's intent must be promulgated and clearly understood two levels down so that commander's intent and the resulting concepts of operation are "nested" to ensure unity of effort. Nested commander's intent ensures that while subordinates have the freedom to conduct their part of the operation as their situation dictates, the results of these disparate actions will contribute to achieving the higher commanders desired end state.

Commander's intent focuses on the enduring portion of any mission—the purpose of the operation—which continues to guide subordinates' actions, while the subordinates' tasks may change as the situation develops. As the commander proceeds through planning and his situational awareness grows, he may refine his intent. He may also include how he envisions achieving a decision—his method—as well as the end state that, when satisfied, accomplishes the purpose of the operation.

The commander's intent provides the overall purpose for accomplishing the task assigned through mission tactics. Although the situation may change, subordinates who clearly understand the purpose and act to accomplish that purpose can adapt to changing circumstances on their own without risking diffusion of effort or loss of tempo. Subordinate commanders will be able to carry on this mission on their own initiative and through lateral coordination with other units.

Commander's Critical Information Requirements

The commander's critical information requirements (CCIRs) identify information on friendly activities, enemy activities, and the environment that the commander deems critical to maintaining situational awareness, planning future activities, and assisting in timely and informed decisionmaking. Commanders use CCIRs to help them confirm their vision of the battlespace, assess desired effects, and how they will achieve a decision to accomplish their mission or to identify significant deviations from that vision.

Not all information requirements support the commander in decisionmaking. CCIRs must be linked to the critical decisions the commander anticipates making. They focus the commander's subordinate commanders and staff's planning and collection efforts. The number of CCIRs must be limited to only those that support the commander's critical decisions. Too many CCIRs diffuse focus.

CCIRs help the commander tailor his command and control organization. They are central to effective information management, which directs the processing, flow, and use of information throughout the force. While the staff can recommend CCIRs, only the commander can approve them. CCIRs are continually reviewed and updated to reflect the commander's concerns and the changing tactical situation.

CCIRs are normally divided into three subcategories: priority intelligence requirements, friendly force information requirements, and essential elements of friendly information. A priority intelligence requirement is an intelligence requirement associated with a decision that will critically affect the overall success of the command's mission. A friendly force information requirement is information the commander needs about friendly forces to develop plans and make effective decisions. Depending on the circumstances, information on unit location, composition, readiness, personnel status, and logistic status could become a friendly force information requirement. An essential element of friendly information is a specific fact about friendly intentions, capabilities, and activities needed by adversaries to plan and execute effective operations against friendly forces.

COMMANDER'S GUIDANCE

Guidance and intent are distinctly different and cannot be used interchangeably. Commander's intent is the purpose of the operation and allows subordinates to exercise judgment and initiative when the unforeseen occurs. Commander's guidance provides preliminary decisions required to focus the planners on the commander's conceptual vision of the operation. The commander develops his commander's initial guidance using how he envisions planning and conducting the operation, his CBAE, his experience, and any information available from higher headquarters. This guidance provides his subordinate commanders and the staff with additional insight of what the force is to do and the resources that will be required to achieve the desired end state. It may be based on the warfighting functions or how the commander envisions the sequence of actions that will allow his force to achieve a decision. The commander may provide general guidance and specific points he wants the staff to consider, like a particular enemy capability, a certain task organization or constraints or restraints from higher headquarters. The commander should articulate those desired effects that will lead to mission accomplishment. This initial guidance is best transmitted to the subordinate commanders and the staff by the commander personally as it will set the direction for the initial planning and preparations and will contribute to establishing tempo in the operation.

Effects are the results of actions—both lethal and nonlethal—taken against the enemy that the commander must achieve to obtain a decision and accomplish his mission. The Marine Corps' warfighting philosophy of maneuver warfare calls for the synchronization of efforts—their arrangement in time, space, and purpose to maximize combat power—to achieve the desired effects on the enemy. Commanders focus combat power to maximize the convergence of effects in time and space at the decisive location and moment. Instead of relying on massed forces and sequential operations, the commander achieves desired effects through the tailored application of the MAGTF's combat power combined with joint combat power. The commander uses lethal and nonlethal means to obtain physical as well as psychological effects against the enemy commander.

There are two forms of effects: direct and indirect. Direct effects are the first-order consequence of a military action, such as the physical attack of an enemy unit or psychological operations to influence the enemy's will to resist. Indirect effects are second- and third-order consequences of military action, often delayed, such as the demoralization of other than the targeted units. Indirect effects can ripple through an enemy force, often influencing other elements of the enemy force physically or psychologically. If combat power is focused on key enemy capabilities and units at the proper place and time, the direct effects of this action may cause the cascade of indirect effects throughout the enemy force and lead to the breakdown of the enemy's will to resist.

Direct and indirect effects often spill over and create unintended consequences, usually in the form of injury or damage to persons or objects unrelated to the objective. Often these unintended consequences can benefit the commander as the effects of friendly actions compound and then cascade throughout the entire enemy force. Integrated planning by the commander includes considering risks of unintended second- and third-order consequences that may result in a negative outcome for friendly forces and noncombatants.

While estimating the outcome of direct and indirect effects can never be an exact process, it becomes increasingly difficult as effects continue to compound and cascade. The commander must consider the factors of METT-T when choosing the desired effect. He must focus on the combination of lethal and nonlethal actions that best accomplishes the desired effects. These effects should be articulated in terms of conditions and measures of effectiveness to facilitate assessment and should be included in the CCIRs. If the desired effect was not achieved, the action may need to be repeated or another method should be sought to achieve the effect.

MISSION

Commanders determine their missions through an analysis of the tasks assigned. This analysis will reveal the essential tasks, together with the purpose of the operation, that clearly indicate the actions required and the desired end state of the operation. The mission includes who, what, when, where, and why the task is to be accomplished.

There are two parts to any mission: the task to be accomplished and the reason or intent behind it. The task describes the action to be taken while the intent describes the purpose of the action. The task denotes what is to be done, and sometimes when and where; the intent explains why. Tasks can be either specified or implied.

Specified tasks are specifically assigned to a unit by its higher headquarters. They are derived primarily from the mission and execution paragraphs of the higher headquarters operation order but may be found elsewhere, such as in the coordinating instructions or annexes.

Implied tasks are not explicitly stated in the higher headquarters order but should be performed to accomplish specified tasks. Implied tasks emerge from analysis of the higher headquarters order, the threat, and the terrain. Routine or continuing tasks are not included in implied tasks.

Essential tasks are those specified or implied tasks that define mission success and apply to the force as a whole. If a task must be successfully completed for the commander to accomplish his purpose, it is an essential task. Once they have been identified as essential tasks, they form the basis of the mission statement.

Tasks normally include a desired end state, which helps the subordinate to more fully understand the purpose of the task and to allow him to make an informed deviation from the stated task if the situation warrants. A task may also have pre-determined conditions that when satisfied tells a subordinate when the desired end state has been reached. End state and conditions help the commander to measure the effectiveness of his subordinate's actions in achieving a decision and accomplishing the mission. See appendix C.

Mission tactics is the assignment of a mission to a subordinate without specifying how the mission must be accomplished. It is a key tenet of maneuver warfare. The higher commander describes the mission and explains its purpose. The subordinate commander determines the tactics needed to accomplish the task based on the mission and the higher commander's intent. Each leader can act quickly as the situation changes without passing information up the chain of command and waiting for orders to come back down.

At the conclusion of mission analysis the commander issues his commander's planning guidance. Planning guidance may be either broad or very detailed depending on the commander and the time and information available. Whatever the nature of this guidance, it must convey the essence of the commander's vision. This guidance should include the commander's vision of decisive and shaping actions and desired effects on the enemy. This assists the staff in applying the elements of the battlefield framework in developing courses of action. The commander's planning guidance may also include phases of the operation, targeting guidance, location and timing of critical events and other aspects of the operations the commander considers important. The commander's planning guidance assists the staff in developing and wargaming courses of action and other planning activities.

Decisive Action

The purpose of all military operations is mission success. Decisive action achieves mission success with the least loss of time, equipment and, most importantly, lives. It causes a favorable change in the situation or causes the threat to change or cease planned and current activities. When a commander seeks battle, he seeks victory: accomplishment of the assigned mission that leads to further significant gains for the force as a whole. Tactical battles are planned for their overall operational and strategic effect. Consequences of a tactical engagement should lead to achieving operational and strategic goals. The goal is not just for the MAGTF commander to achieve a decision, but to ensure that decision has greater meaning by contributing to the success of his senior commander's operation or campaign. For an action to be truly decisive, it must lead to a result larger than the action itself. Decisive action creates an environment where the enemy has either lost the physical capability or his will to resist. Forcing the enemy to reach a culminating point could be a decisive action by a defending force. A culminating point is that point in time and space where an attacker's combat power no longer exceeds that of the defender and/ or an attacker's momentum can no longer be sustained. A culminating point for a defender is that point in time when a defender must withdraw to preserve his force.

Decisive action may occur anywhere and at any time in the single battle. Any of the MAGTF's three major subordinate commands can achieve a decision. The ACE and the GCE normally achieve a decision through combat. The CSSE may be called upon to achieve a decision in MOOTW; e.g., humanitarian assistance. The commander considers the following in planning decisive action:

- What are the enemy's intentions?
- What are the centers of gravity and critical vulnerabilities?
- What is the battlespace and is it appropriate to the MAGTF's capabilities?
- What are the effects necessary to achieve a decision?
- Are the command relationships appropriate to the mission and battlespace?
- Have proper missions been assigned to the main effort, supporting efforts, and the reserve?
- How to synchronize the actions of the major subordinate commands?
- Does the MAGTF have the resources to accomplish the mission within the battlespace assigned?
- Have the MAGTF's resources been allocated and apportioned properly?
- Can the MAGTF be sustained in its effort to achieve a decision?

- Can the MAGTF accomplish the mission without reaching a culminating point?
- How does the MAGTF commander recognize whether the MAGTF succeeded in executing the plan?
- How does the commander assess success and whether changes must be made?

Decisive action at the MAGTF level involves more than just fire and maneuver. The MAGTF commander arranges a series of battles or engagements to achieve a decision. The commander arranges the actions of the MAGTF in terms of time, space, and resources to generate combat power at the decisive time and place.

Shaping Actions

The MAGTF commander sets the conditions for decisive action by conducting shaping actions to achieve desired effects. Shaping is all lethal and nonlethal activities conducted throughout the battlespace to influence a threat capability, force or the enemy commander's decision. The commander shapes the battlespace principally by creating conditions to protect friendly critical vulnerabilities and attack enemy critical vulnerabilities. In many cases, the MAGTF can achieve much of its own shaping. The objective of shaping actions might include—

- Limit enemy freedom of action.
- Deny the enemy the capability to concentrate forces.
- Deceive the enemy as to friendly intentions.
- Destroy enemy capabilities.
- Alter the tempo of operations.
- Gain and maintain momentum.
- Influence perceptions of the enemy, allies, and noncombatants.

Shaping can have a favorable impact on friendly forces. The sense of being on the offensive and taking the fight to the enemy helps to maintain morale and foster offensive spirit for later decisive actions. Shaping incorporates a wide array of functions and capabilities to achieve desired effects and is more than just fires and targeting. It may include, but is not limited to, direct attack, psychological operations, electronic warfare, deception, civil affairs, information management, public affairs, engineer operations, and preventive medical services. Logistics operations, such as the marshalling of critical ammunition, fuel, and supplies to facilitate future operations, shape both friendly and threat forces.

Shaping makes the enemy vulnerable to attack, impedes or diverts his attempts to maneuver, aids the MAGTF's maneuver, and otherwise dictates the time and place for decisive action. It forces the enemy to abandon their course of action and adopt a course of action favorable to the MAGTF. Shaping actions must be relevant to the envisioned decisive action. The commander attempts to shape events in a way that allows him several options, so that by the time the moment for decisive action arrives, he is not restricted to only one course of action. The goal of shaping is to eliminate the enemy's capability to fight effectively before the MAGTF initiates decisive action. As stated in MCDP 5, ideally, when the decisive moment arrives, the issue has been resolved. Our actions leading to this point have so shaped the conditions that the result is a matter of course.

BATTLEFIELD FRAMEWORK

This framework describes how the commander will organize his battlespace and his forces to achieve a decision. The battlefield framework consists of the battlespace organization of envisioned deep, close, and rear tactical operations as well as the organization of the force into the main effort, reserve, and security. Supporting efforts are addressed in the context of deep, close, and rear operations as part of the single battle. The battlefield framework provides the commander and his staff with an organized way to ensure that they consider in planning and execution all essential elements of successful military operations.

INTEGRATED PLANNING

Integrated planning is a disciplined approach that is systematic, coordinated, and thorough. It uses the warfighting functions to integrate the planning and supervise execution. Planners use integrated planning to consider all relevant factors, reduce omissions, and share information across the warfighting functions.

Integrated planning is essential to eliminate "stove pipe" planning when individual planners, staff sections, and functional areas plan in a vacuum, without coordination with others. This approach often results in disjointed plans and execution that is not synchronized. Staffs will produce more useful operation plans and orders and commanders will realize more synchronized operations across the elements of the MAGTF with increased tempo.

The warfighting functions are used extensively in integrated planning. Commanders and staffs use warfighting functions as a planning framework. Their use ensures that the commander and his planners consider all critical functional areas when planning and making decisions. Warfighting functions are planning and execution tools used by planners and subject matter experts in each of the functional areas to produce comprehensive plans that are integrated with the other warfighting functions. This integration of the planning effort helps the commander to achieve unity of effort.

SINGLE BATTLE

The MAGTF commander conducts operations within the context of the single battle. Single battle allows the commander to effectively focus the efforts of all MAGTF elements of the force to accomplish his mission. Within the single battle, the commander conducts centralized planning while fostering decentralized execution allowing subordinates to exercise disciplined initiative and exploit opportunities. Centralized planning is essential for controlling and coordinating the efforts of all available forces. Decentralized execution is essential to generate the tempo of operations required and to cope with the uncertainty, disorder, and fluidity of combat.

A commander must always view his AO as an indivisible entity. Operations or events in one part of the AO may have profound and often unintended effects on other areas and events. While the AO may be conceptually divided to assist centralized planning and decentralized execution, the commander's intent ensures unity of effort by fighting a single battle. The asymmetrical nature of the MAGTF elements makes this particularly critical to the success of the MAGTF's operations. See figure 6-4.

Under single battle, the AO consists of three major areas—deep, close, and rear—where distinctly different operations are performed. These operations are not necessarily restricted to or characterized by distance or location in the AO. They are functional actions that must be accomplished for other functions to be effective. The MAGTF does not merely divide the battlespace up with the ACE taking the deep, the GCE taking the close, and the CSSE taking the rear area. The MAGTF commander is in charge and is responsible for the entire battle. To synchronize actions within the single battle, the commander must determine what, where, when, and how to apply the warfighting functions.

While the MAGTF commander desires to defeat the enemy in a single battle or engagement, it may be beyond the capabilities of the MAGTF to achieve this. Thus, MAGTF operations may need to be phased. All actions and phases must be connected and focused on achieving a decision. This arrangement of forces in time and space to generate sufficient combat power to achieve a decision is the result of detailed and integrated planning.

Deep Operations

Deep operations shape the battlespace to influence future operations. They seek to create windows of opportunity for decisive action, restrict the enemy's freedom of action, and disrupt the cohesion and tempo of his operations. Deep operations help the commander seize the initiative and set the conditions for close operations. Because of its operational reach, deep operations are primarily conducted by the ACE, although the GCE and CSSE may play significant roles. MAGTF intelligence assets; e.g., force reconnaissance and signals intelligence and ACE and GCE surveillance and reconnaissance assets (UAVs and ground surveillance radars) contribute to the conduct of deep operations.

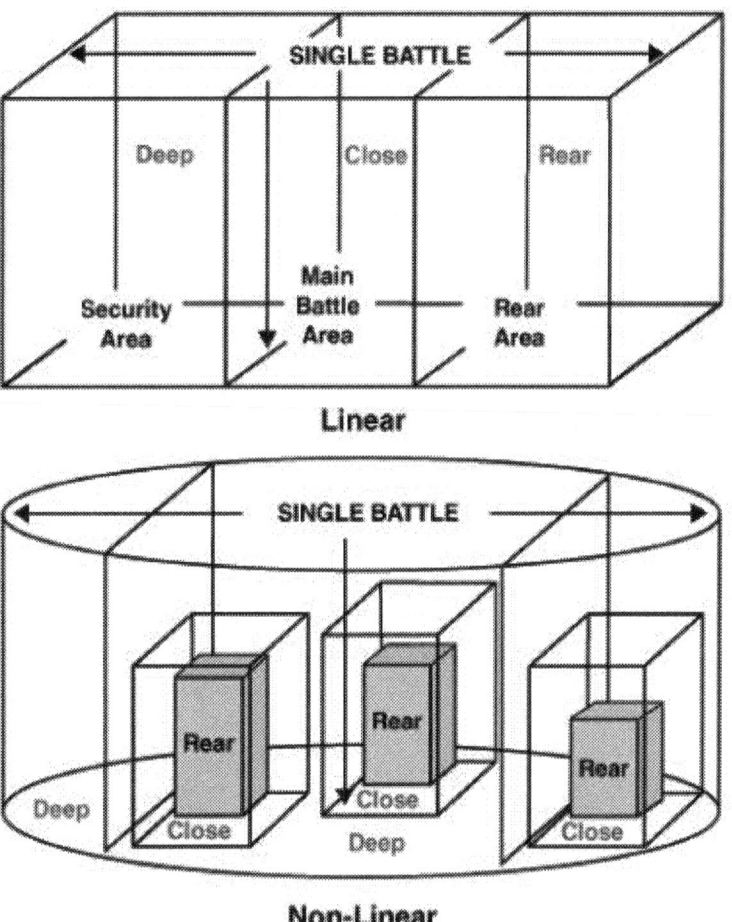

Figure 6-4. Single Battle.

The commander focuses on attacking enemy capabilities—moral and physical—that most directly contribute to the accomplishment of his mission. Deep operations should exploit or create these critical enemy vulnerabilities. Deep operations normally focus on the enemy's follow-on and supporting forces, command and control nodes, and key lines of communications or facilities. Deep operations may require coordination and integration with national-level assets and joint forces. They may include—

- Interdiction through fires and maneuver.
- Surveillance, reconnaissance, and target acquisition.
- IO such as deception or psychological operations.
- Offensive antiair warfare.

Close Operations

Close operations project power against enemy forces in immediate contact and are often the decisive actions. These operations require speed and mobility to rapidly concentrate overwhelming combat power at the critical time and place and exploit success. Close operations are dominated by fire and maneuver conducted by combined arms forces from the GCE and the ACE. Combined arms forces maneuver to enhance the effects of their fires and fire to enhance their ability to maneuver. As they maneuver to gain positions of advantage over the enemy, combined arms forces deliver fires to disrupt the enemy's ability to interfere with that maneuver. Commanders prioritize fires to weight the main effort and to focus combat power to achieve effects that lead to a decision. The effects of fires can be massed to strike the enemy at the decisive point and time, while reducing the risks to the force entailed in massing maneuver forces at a single point or in a single portion of the battlespace.

Rear Operations

Rear operations support deep and close operations and facilitate future operations. Security is inherent in rear operations. Sustainment must not be interrupted and assets must be protected. Rear operations ensure the freedom of action of the force and provide continuity of operations, logistics, and command and control. Rear area operations deny the use of the rear area to the enemy. To minimize the logistical footprint, rear operations may require the maximum use of sea-basing, push logistics, host-nation support, and existing infrastructure. Rear operations are conducted by all MAGTF elements.

Rear area operations are evolutionary in nature. As the operation progresses, the geographic location, command and control structure, and the organization of the rear area can be expected to change. The broad functions of rear area operations, as delineated in joint and Marine Corps doctrine, include—

- Communications.
- Intelligence.
- Sustainment.
- Security.
- Movement.
- Area management.
- Infrastructure development.
- Host-nation support.

To provide command and control of rear area operations, the commander may assign a rear area coordinator or commander with specific, designated functions. He usually establishes a rear area operations center to assist in the conduct and coordination of those functions of rear area operations assigned. For more information, see MCWP 3-41.1, *Rear Area Operations*.

Noncontiguous and Contiguous

The battlefield framework may reflect linear operations where there is a continuous and contiguous array of units across the AO and through the depth of the deep, close, and rear areas. A more likely situation is one where the MAGTF conducts nonlinear operations within a noncontiguous battlespace and within an operational framework with noncontiguous deep, close, and rear areas. Operation Restore Hope in Somalia (1992–1993) is an example of a battlefield framework with noncontiguous areas. The MAGTF's rear area was centered around the separate sites of the embassy compound, port, and airfield in the city of Mogadishu, while its close area was widely scattered around the towns and villages of the interior that were occupied by the MAGTF. The MAGTF's deep area included the rest of the country and particularly those population and relief centers not under the joint force commander's supervision.

The MAGTF commander must be well versed in the capabilities and limitations of his forces and their role in deep, close, and rear operations to conduct the single battle. He must consider that there may be deep, close, and rear operations at every level of command. For example, a subordinate commander's deep operations may constitute part of the higher commander's close operations. See figure 6-5 on page 6-24.

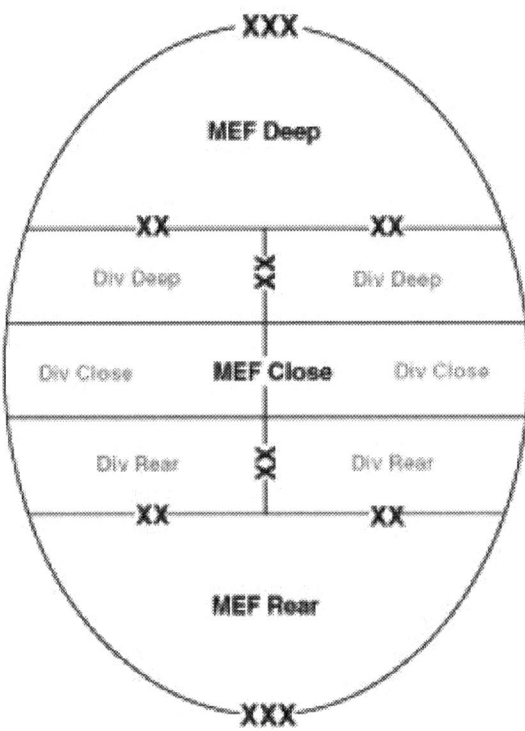

Figure 6-5. Battlespace Organization.

By conceptually dividing the AO and using the warfighting functions to conduct integrated planning for each area, the commander ensures the coordination of his forces in executing the single battle. It is important to remember that the enemy's disposition and actions will seldom coincide with how the Marine commander has organized his AO. Therefore, the commander's planning and execution must be flexible enough to accommodate this difference and exploit resulting opportunities.

MAIN AND SUPPORTING EFFORTS

The main effort is a central concept of maneuver warfare. It calls for concentrating efforts on achieving objectives that lead to victory. The main effort is that unit assigned to accomplish the mission or task critical to mission success. *The main effort normally is that unit with which the commander plans to conduct the decisive action and it should be selected, reinforced, and supported accordingly.* The commander assigns the main effort to a specifically designated subordinate unit.

The commander focuses the combat power of the force against enemy critical vulnerabilities in a bold bid to achieve decisive results. The main effort may be viewed as a harmonizing force for subordinate's initiative.

The main effort may be from any MAGTF element or force assigned. MCDP 1-3 says the commander provides the bulk of his combat power or other assets to the main effort to maintain momentum and ensure accomplishment of the mission. These assets may include not only maneuver forces but also capabilities that enhance the main effort's ability to accomplish its mission. The commander normally gives the main effort priority of various types of support. It is also provided with the greatest mobility and the preponderance of combat support and combat service support. However, overburdening the main effort with unnecessary assets can degrade its ability to move rapidly and decisively. The reserve is positioned to best exploit the main effort's success.

The commander may concentrate the combat power of the main effort by assigning it a narrower zone of action or reducing its AO. In summary, the commander weights the main effort by task-organizing his force or by providing priority of—

- Air support.
- Fire support assets.
- Transportation and mobility assets such as heavy equipment transporters, assault support helicopters, bridging, and obstacle clearing engineer support.
- Combat service support to preclude the main effort from reaching the culminating point prematurely. This support might include mobile CSSEs, critical supplies like fuel and ammunition, and exchange or rapid repair of essential equipment.
- Specialized units or capabilities, such as civil affairs or psychological operations units during MOOTW.
- Personnel replacements.
- Command and control support.
- Intelligence support.

The commander disguises the main effort until it is too late for the enemy to react to it in strength. He accomplishes this through demonstrations and feints, security, cover and concealment, and by dispersing his forces until the last instant and achieving mass at the critical time and place.

Supporting efforts help shape the battlespace in support of the main effort's envisioned decisive action. Faced with a decision, commanders of supporting efforts must ask themselves how can I best support the main effort? Conversely, they must avoid actions that do not contribute to the success of the main effort. There may be more than one supporting effort in support of the main effort.

Massing of combat power to support the main effort may require time and additional transportation assets to marshal the necessary support. Task organization of supporting units may also be required to provide responsive and flexible support.

Commanders apply the principle of economy of force to supporting efforts. They make effective use of available assets needed to support the main effort, while conserving others for future actions. MCDP 1-3 states that forces not in a position to directly support the main effort should be used to indirectly support it. For example, a commander may use other forces to deceive the enemy as to the location of the main effort. Such forces might be used to distract the enemy or to tie down enemy forces that might otherwise reinforce the threatened point. The commander weighs the value of the deception against the cost in terms of forces and assets needed to portray a credible force. Uncommitted forces can be used in this effort by maneuvering them in feints and demonstrations that keep the enemy off balance.

The reserve may be tasked to support the main effort and often will become the main effort when employed. It is important for the commander to ensure that the reserve is not assigned nonessential tasks that degrade its ability to respond rapidly to fleeting opportunities created by the main effort or to reinforce the main effort at the decisive time and place.

While a commander always designates a main effort, it may shift—either planned or unplanned—during the course of a battle as events unfold. Because events and the enemy are unpredictable, few battles flow exactly as the commander has planned. A supporting effort may achieve unexpected success during execution. As a result, the commander must make adjustments. After assessing the changing situation, he may designate the supporting effort as the main effort. Commanders of supporting efforts must be prepared to assume the role of the main effort as the situation changes as a result of emerging opportunities or unforeseen setbacks.

There may be costs in shifting the main effort. The larger the organization, the more costly this shift may be. The costs include the time and effort to shift resources and priority of support (fires, supply, transportation, medical, engineering). The commander must weigh the benefits and costs for shifting the main effort. He should only shift the main effort when he is convinced this will lead to decisive action.

The flexibility inherent in Marine aviation allows the commander to shift the main effort to the ACE rapidly and usually with significantly less repositioning of resources than other forces of the MAGTF. Normally the greatest impact on the ACE is that the new tasks resulting from the shift of the main effort may not be performed as timely or with the optimum combination of forces and ordnance as the previously planned tasks. Forces may be required to reorganize and plan new missions or to redirect previously planned missions designed to attack specific targets against new targets.

To conduct decisive actions and to weight the main effort sufficiently so that he can achieve a decision, the commander must organize his assigned and attached forces for specific missions and tasks. This process of allocating available assets to subordinate commanders and establishing appropriate command and supporting relations is called task-organizing. The grouping of forces or units to accomplish a specific mission or task is task organization. Marine Corps forces are task-organized routinely and are used to operate in task forces to accomplish specific missions and then rapidly resume their duties with their parent command. While taking advantage of the close coordination and cooperation realized by units with habitual relations with other units; e.g., Marine artillery units habitually support certain infantry regiments, these forces are agile enough to assume a new supporting relationship or attachment to a different unit.

The ability to rapidly tailor Marine Corps forces through task organization to accomplish a wide array of missions or tasks allows the commander to effectively and efficiently use the forces and assets available to him. It is incumbent on the commander to understand the capabilities and limitations of the forces available to develop the best possible task organizations. He must also realize that the creation of a task-organized force will take some time and may have an affect on his forces' tempo. Frequent or gratuitous task-organizing may actually reduce the effectiveness of the force.

THE RESERVE

The reserve is an essential tool used by the commander to exploit success. The reserve is part of the commander's combat power initially withheld from action in order to influence future action and deal with emerging opportunities or a crisis. The reserve provides the commander the flexibility to react to unforeseen developments. Often a commander's most difficult and important decision concerns the time, place, and circumstances for committing the reserve. While the commander sometimes must employ his reserve to deal with a crisis, he should always attempt to use the reserve to reinforce success and exploit opportunities to achieve a decision. The commander uses his reserve to restore momentum to a stalled attack, defeat enemy counterattacks, and exploit success. Once committed, the reserve's actions normally become the decisive operation. Every effort is made to reconstitute another reserve from units made available by the revised situation. Since the reserve is often the commander's bid to achieve a decision, it is usually designated the main effort when committed. The reserve is not to be used as a follow and support force or a follow and assume force.

The reserve should be as strong a force as possible with appropriate mobility and firepower. Its strength and location will vary with its contemplated mission, the form of maneuver, the terrain, the possible enemy reaction, and the clarity of the situation. The commander should organize, equip, and rehearse the reserve for the intended mission. He should not constitute his reserve by weakening his decisive operation. A reserve must have mobility equal to or greater than the most dangerous enemy threat, and it must be able to fight the most dangerous enemy threat. The more uncertain the situation, the larger should be the reserve. When the situation is obscure, the reserve may consist initially of the bulk of the force, centrally located and prepared to be employed at any point. The commander only needs to provide a small reserve to respond to unanticipated enemy reactions when he has detailed information about the enemy. However, the reserve must always be sufficient to exploit success.

The commander must also consider intangible factors when selecting and tasking a reserve, including the proficiency, leadership, morale, fatigue and combat losses, and maintenance and supply status of the unit. Care is taken in the positioning of the reserve to balance force protection requirements with the imperative to best position the reserve to enhance its ability to exploit opportunities. When committed, the reserve—as the main effort—receives priority for resources and services.

SECURITY

Security is inherent in all MATGF operations and includes those measures taken by a military unit, an activity, or installation to protect itself against all acts designed to, or which may, impair its effectiveness. See chapter 11. Security operations are an element of overall force protection measures that must be conducted in all operations, whether offense or defense. Sound security operations are based on—

- **Orientation.** Security forces position themselves between the main force and the enemy. Security elements depend on the movement of the main force. The operations of the security force must be closely coordinated with the concept of operations.

- **Reconnaissance.** Security forces conduct reconnaissance, which seeks to reduce unknowns for the commander. The security force reduces the chance of surprise to friendly forces.
- **Early and Accurate Warning.** Unless warning of an enemy threat reaches the commander in time for him to react, the warning is useless. An erroneous warning may be as detrimental as no warning as all.
- **Reaction Time and Maneuver Space.** The security force gives the commander the time and space to counteract an enemy threat. A security force executes its mission to the greatest depth possible based on its capabilities and the tactical situation.
- **Gain and Maintain Contact.** Security forces seek to gain contact with the enemy as early as possible. Based on the assigned mission and the capabilities of the security force, contact may vary from observation to combat. Contact is normally not broken off without permission from higher headquarters because accurate information about the enemy's location, disposition, and movement prevents surprise by the enemy. However, the requirement to maintain contact must be balanced with the friendly concept of operations.
- **Conduct Counterreconnaissance.** Counterreconnaissance is an important aspect of all security operations. It is all measures taken to prevent hostile observation of a force, area or place.
- **Mobility.** Security forces normally require mobility equal or greater than that of the enemy. In mountains, dense forests or built up urban areas, dismounted lighter forces may have greater relative mobility than mechanized forces. The inherent mobility of the ACE provides the MAGTF with an ideal force for the conduct of selected security operations.

PHASING

Phasing assists the commander and staff in planning and executing operations. The commander uses phasing to divide his vision for how he intends to accomplish his mission into portions that reflect the requirement to perform a major task to achieve a decision. A change in phase usually involves a change of a major task. I MEF offensive operations during Operation Desert Storm (1991) was conducted using the following phases:

- **Phase I.** Strategic air operations to attain air supremacy, attack Iraqi warmaking infrastructure, and destroy the Republican Guard.
- **Phase II.** Attainment of air supremacy in the Kuwait theater of operations and the suppression of the Iraqi integrated air defense system.

- **Phase III.** Preparation of the battlespace to reduce the combat effectiveness of the enemy in the Kuwait theater of operations, specifically the reduction of Iraqi armor and artillery by 50 percent and FROG (free rocket over ground) missiles and multiple rocket launchers by 100 percent.
- **Phase IV.** Ground offensive operations.

Commanders should establish clear conditions for the initiation and termination of each phase. While phases are distinguishable by friendly forces, they should not be readily apparent to the enemy. The commander should take whatever actions necessary to conceal from the enemy the distinctions and, most especially, the transitions between phases.

Phases may be further subdivided into stages to provide greater detail in planning and enhance control and coordination in execution. I MEF's Phase IV ground offensive operations consisted of—

- **Stage A.** Penetration of forward Iraqi defenses.
- **Stage B.** Exploit the success of the penetration to destroy Iraqi forces in zone.
- **Stage C.** Consolidate to prevent reinforcement or escape of Iraqi forces in zone.

Phases and stages should be aligned as closely as possible with those adopted by higher headquarters to reduce confusion during transition from one phase or stage to another and to enhance coordinated efforts.

OPERATION PLANS AND ORDERS

Operation plans and orders communicate the commander's intent, guidance, and decisions in a clear, useful, and timely form. They should be easily understood by those who must execute the order. Operation plans and orders should only contain critical or new information—not routine information and procedures normally found in standing operating procedures.

Plans and orders should be the product of integrated planning to eliminate stovepiping of information. Critical information, such as the mission, commander's intent, and tactical tasking should be prominently positioned in the basic order. Concepts of maneuver, fires, and support should also be in the basic order. Planners need to reconcile plans and orders to ensure all critical information is presented. Comparison or "crosswalking" of the plan or order with those of higher, adjacent, and subordinate commands help to identify conflicts or omissions and achieve unity of effort. The operation plan or order is the source of authority on the operation, *not* briefing slides and e-mail messages. Chiefs of staff must maintain close control over versions and changes to the plan or order to ensure unauthorized changes do not get promulgated and that the approved version is properly disseminated.

TRANSITIONING BETWEEN PLANNING AND EXECUTION

The actions of the commander and the staff during the transition from planning to execution may be of critical importance in accomplishing the mission. Transition ensures a successful shift from planning to execution. It enhances the situational awareness of those who must execute the plan, maintains the intent of the concept of operations, promotes unity of effort, and generates tempo. Transition facilitates the synchronization of plans between higher and subordinate commands and aids in integrated planning by ensuring the synchronization of the warfighting functions. At the MAGTF level (where the planners may not be the executors), transition provides a full understanding of the plan to those who were not involved in its development.

Transition occurs at all levels of command. A formal transition normally occurs on staffs with separate planning and execution teams. Planning time and personnel may be limited at lower levels of command, such as the regiment, aircraft group, or below. Therefore, transition may take place intuitively as the planners are also the executors. Transition may be accomplished through the assignment of a plan proponent—a planner who aids the executors in interpreting and applying the plan in action—and through participation in transition briefs, drills, and a confirmation brief. Confirmation briefs are a particularly valuable technique to ensure synchronization or nesting of higher, adjacent, and subordinate command's plans.

EXECUTION

Execution of MAGTF operations is the concerted action of the commander and his forces to conduct operations based on the operation plan or order, modified as the current tactical situation dictates, to achieve the commander's end state and accomplish the mission. The commander and his forces must seize and retain the initiative, create overwhelming tempo, establish and maintain momentum, achieve desired effects, exploit success, and successfully finish the operation. He commands the activities of his various subordinate units and assesses the success of those activities in obtaining the goals of the operation.

The commander, assisted by his deputy commander and the staff, must control or coordinate the activities of all of the MAGTF elements. These activities include the movement and maneuver of the force, coordination and control of fires, collection of intelligence, sustainment and protection of the force, and assessment of these activities to determine the progress of the command in achieving the desired end state. He must command the force and supervise the activities of his subordinate commanders in carrying out his mission and intent.

Command and Control

MCDP 6 states that no single activity in war is more important than command and control. It is the means by which the commander recognizes what needs to be done and sees to it that appropriate actions are taken. Command and control provides purpose and direction to the varied activities of a military unit. If done well, command and control adds to the strength of the force—if done poorly, it may be a liability to the force.

JP 1-02 defines command as the authority that a commander in the Armed Forces lawfully exercises over subordinates by virtue of rank or assignment. Command includes the authority and responsibility for effectively using available resources and for planning the employment of, organizing, directing, coordinating, and controlling military forces for the accomplishment of assigned missions. The commander must effectively command the activities of his subordinate commanders during operations. His span of control should not exceed his capability to effectively command.

Command in battle incorporates two vital skills—the ability to decide and the ability to lead. They integrate a commander's vision of the situation and battlespace and how he plans to achieve his desired end state with leading, guiding, and motivating subordinates. These two skills are tightly interwoven and are the central factors from which the warfighting functions are integrated to create combat power and conduct expeditionary maneuver warfare.

Leadership is the influencing of people to work toward the accomplishment of a common objective and is essential to effective command. While the component, MAGTF, and major subordinate command commanders exercise leadership by visualizing and describing how the operation will be conducted, commanders at lower levels accomplish the goals of the operation by motivating and directing the actions of their units.

The ability to command and control an organization is enhanced when the commander decentralizes decisionmaking authority as much as each situation allows. This means that commanders on the scene and closest to the events have the latitude to deal with the situation as required *on their own authority*—but always in accordance with the higher commander's intent. Decentralization speeds up reaction time: the commander does not have to wait for information to flow up to a higher commander and orders to flow back down. Confidence in the abilities of subordinates is an important part in decentralization. Leaders who have confidence in their subordinates will feel more comfortable in granting them greater latitude in accomplishing tasks. It fosters a climate where senior leaders know that their intent will be carried out.

Control can generally be divided into two types: centralized and decentralized. Centralized control tends to work from the top down: the commander determines what his subordinates will and will not do. Decentralized control works from the bottom up. While cooperation is required for both types of control, it is essential in decentralized control. Subordinates work together laterally and from the bottom up to accomplish tasks that fulfill the commander's intent.

Battle rhythm is an important aspect of command and control. The commander must ensure that the planning, decision, and operating cycles of his command are nested or linked to that of his higher headquarters and that his subordinate commanders synchronize their battle rhythms with his headquarters. Ensuring information and requests for support are forwarded to higher headquarters in time for that headquarters to act increases the likelihood that the command will obtain the desired support or effects. Some of the planning, decision, and operating cycles that influence the battle rhythm of the command include the intelligence collection cycle, targeting cycle, air tasking order cycle, reconnaissance tasking cycle, and the battle damage assessment collection cycle.

Effective decisionmaking is essential to command and control. Commanders develop information management processes to ensure access to timely and useful information to make decisions. Information management is the processes and techniques the command uses to obtain, manipulate, direct, control, and safeguard information. Sound information management practices facilitates the rapid, distributed, and unconstrained flow of information in all directions—to higher headquarters, adjacent units, and subordinate commanders. Information management policies and procedures enable the staff to determine the importance, quality, and timeliness of information to provide the commander with focused information to prevent information overload.

Assessment

Assessment is the continuous appraisal of military operations to determine progress toward established goals. It answers the commander's question *"How are we doing?"* It helps the commander recognize whether his planned activities are achieving their desired effects and whether he has to modify or cease those activities to achieve his desired end state. Assessment is continuous and is focused on the overall effectiveness of the command in achieving the commander's goals. Assessment is the basis for the commander's decisions concerning future actions. It allows him to rapidly act to exploit unexpected success or opportunity and to counter unanticipated enemy success.

Successful assessment requires a commander who can clearly and accurately visualize the battlespace and the operation. It requires situational awareness on the part of the commander that allows him to recognize the difference between the desired effects and the actual effects of the operation. This perceived difference between what was planned and what actually happened then becomes the catalyst for decisionmaking.

Commanders assess their operation's effectiveness by measuring how successful they have been in completing the tasks stated or inherent in their mission. They determine if operations have met the conditions previously established that support an upcoming decision by the commander or if the task has been completed. Conditions should be linked to the purpose of the task and be understandable, relevant, and measurable. Since some conditions are necessarily complex, commanders and their staffs may also use measures of effectiveness to further describe those conditions that must be met before a task is completed or a new phase of the operation can commence. Measures of effectiveness are indicators that demonstrate the degree to which a condition has been satisfied. They provide the commander with a tangible indicator of how close he is to achieving his desired conditions.

The intelligence collection effort, as well as the overall combat reporting process in the force, must focus on providing timely and useful information to the commander to aid him in his assessment of operations. The fulfillment of CCIRs and priority intelligence requirements will often be critical in determining whether the task has been completed and the conditions exist to support transition from one phase of the operation to another. While assessment routinely takes place throughout the planning, deployment, and redeployment phases of an operation it is truly essential during execution.

TACTICAL TENETS

Actions at the tactical level of war are the building blocks the MAGTF commander uses to achieve operational success and fulfill the joint force commander's operational goals. Every action the MAGTF commander and the major subordinate element commanders take is aimed at achieving the senior commander's goals and accomplishing their mission. The tactical level of war is the province of combat. It includes the maneuver of forces to gain a fighting advantage; the use and coordination of fires; the sustainment of forces throughout combat; the immediate exploitation of success; and the combination of different arms and weapons—all to cause the enemy's defeat. It is the MAGTF that conducts these tactical operations through the major subordinate commands that execute these tactical operations. See appendix D.

Successful execution of Marine Corps tactics requires the thoughtful application of a number of tactical tenets to succeed on the battlefield. Key among them are achieving a decision, gaining advantage, tempo, adapting, and exploiting success and finishing. They do not stand alone but are combined to achieve an effect greater than their separate sum. Part of the art of tactics is knowing where and when to apply these tenets and how to combine them to achieve the desired effect.

Achieving a Decision

The objective of tactics is to achieve military success through a decision in battle. In combat, the success the commander seeks is victory—not a partial or marginal outcome, but a victory that settles the issue in his favor and contributes to the success of the overall campaign.

Achieving a decision is *not easy*. The enemy's skill and determination may prevent even a victorious commander from achieving the decision he seeks. Commanders must not engage in battle without envisioning a larger result for their actions.

Perfect understanding of the situation or a highly detailed plan is useless if the commander is not prepared to act decisively. When the opportunity arrives, the commander must exploit it fully and aggressively, committing every ounce of combat power he can muster and pushing the force to the limits of exhaustion. Key to this effort is identifying enemy critical vulnerabilities, shaping the operating area to gain an advantage, designating a main effort to focus the MAGTF's combat power, and acting in a bold and relentless manner.

Forcing a successful decision requires the commander to be bold and relentless. Boldness refers to daring and aggressiveness in behavior. It is one of the basic requirements for achieving clear-cut outcomes. The commander must have a desire to "win big," even if he realizes that in many situations the conditions for victory may not yet be present. Relentlessness refers to pursuing the established goal single-mindedly and without let up. Once he has an advantage, he should exploit it to the fullest. He should not ease up, but instead increase the pressure. Victory in combat is rarely the product of the initial plan, but rather of relentlessly exploiting any advantage, no matter how small, until it succeeds.

Gaining Advantage

Combat is a test of wills where the object is to win. One way to win is to gain and exploit every possible advantage. The commander uses maneuver and surprise whenever possible. He employs complementary forces as combined arms. He exploits the terrain, weather, and times of darkness to his advantage. He traps the enemy by fires and maneuver. He fights asymmetrically to gain added advantage. He strives to gain an advantage over the enemy by exploiting every aspect of a situation to achieve victory, not by overpowering the enemy's strength with his own strength.

Combined Arms

Combined arms is a Marine Corps core competency. The use of combined arms is a key means of gaining advantage. Combined arms presents the enemy not merely with a problem, but with a dilemma—a no-win situation. The commander combines supporting arms, organic fires, and maneuver in such a way that any action the enemy takes to avoid one threat makes him more vulnerable to another.

Modern tactics is combined arms tactics. It combines the effects of various arms—infantry, armor, artillery, and aviation—to achieve the greatest possible effect against the enemy. The strengths of the arms complement and reinforce each other. At the same time, weaknesses and vulnerabilities of each arm are protected or offset by the capabilities of the other.

The MAGTF is a perfect example of a balanced combined arms team. For example, an entrenched enemy discovers that if he stays in fighting holes, Marine infantry will close with and destroy him. If he comes out to attack, Marine artillery and aviation will blast him. If he tries to retreat, Marine mechanized and aviation forces will pursue him to his destruction. Combined arms tactics is standard practice and second nature for all Marines.

Exploiting the Environment

The use of the environment offers tremendous opportunities to gain advantage over the enemy. Marines must train for and understand the characteristics of any environment where they may have to operate: jungle, desert, mountain, arctic, riverine or urban. More importantly, Marines must understand how the terrain, weather, and periods of darkness or reduced visibility impact on their own and the enemy's ability to fight.

In addition to the physical aspects of the environment, Marines must consider the impact on the operation by the people and the culture, political and social organization, and any external agencies or organizations that exist within the AO.

As most expeditionary operations take place in the world's littoral regions that are more densely populated and contain more urban areas than the hinterlands, Marines must plan for and be prepared to conduct more civil-military operations.

Complementary Forces

Complementary forces—the idea of fix-and-flank—are an important way of gaining advantage. The commander seeks to crush the enemy—as between a hammer and an anvil—with two or more actions. With its two combat arms, the MAGTF has organic complementary and asymmetric forces. Ground combat forces may attack an enemy in one direction and dimension, while aviation combat forces are attacking from another direction and dimension. This capability places the enemy in a dilemma. The opponent is now vulnerable to one or the other of the two combat forces. He has no protection against both-no matter how he moves he is exposed.

One of the complementary forces may take a direct, obvious action to fix the enemy. The other force takes the unexpected or extraordinary action. These two actions work together against the enemy. The two actions are inseparable and can be interchangeable in battle. The concept is basic, but it can be implemented in a variety of combinations limited only by imagination.

Surprise

Achieving surprise can greatly increase advantage. In fact, surprise can often prove decisive. The commander tries to achieve surprise through information operations that result in deception, stealth, and ambiguity.

The commander uses deception to mislead the enemy with regard to his real intentions and capabilities. By employing deception, he tries to cause the enemy to act in ways that will eventually prove prejudicial for them. He may use deception to mislead the enemy as to the time and location of a pending attack. He may use deception to create the impression that his forces are larger than they really are. Forces used to support deception operations must be appropriate to and of sufficient size to make the deception credible. The commander hopes the enemy will realize this deception only when it is too late for them to react.

Surprise can be generated through stealth. Stealth is used to advantage when maneuvering against an enemy. It provides less chance of detection by the enemy, leaving him vulnerable to surprise action for which he may be unprepared.

The commander can also achieve surprise through ambiguity. It is usually difficult to conceal all friendly movements from the enemy, but the commander can sometimes confuse him as to the meaning of what he sees, especially his awareness of where the main effort is or where the commander has placed his bid for a decision. Clearly, the ambiguity created in the minds of the Iraqi high command by the amphibious force contributed to fixing the Iraqi forces, allowing coalition forces to succeed in Operation Desert Storm.

Asymmetry

Fighting asymmetrically means gaining advantage through imbalance, applying strength against an enemy weakness in an unexpected way. At the tactical level, fighting asymmetrically uses dissimilar techniques and capabilities to maximize the MAGTF's strengths while exploiting enemy weaknesses. By fighting asymmetrically, the MAGTF does not have to be numerically superior to defeat the enemy. It only has to be able to exploit the enemy's vulnerabilities in all dimensions. In MAGTF operations, using tanks to fight enemy artillery or attack helicopters against enemy tanks are examples of fighting asymmetrically. In these examples, the tanks' and aircraft's greater speed and mobility provides an advantage over the enemy. Fast moving tanks operating in the enemy's rear against stationary or slow moving artillery can disrupt the enemy's cohesion. Ambushing tanks with attack helicopters in terrain that hampers tank maneuver provides even greater effect and generates even more advantage. United States attack helicopters assisted in blunting a rapid and powerful North Vietnamese advance and destroyed the enemy's armor using just such tactics during the Easter offensive of 1972 in the Republic of Vietnam.

Commanders must anticipate asymmetric actions and take measures to reduce the adversary's advantage. An adversary may counter MAGTF strengths asymmetrically by conducting insurgency or terrorist operations, such as the terrorist bombing of the Marine barracks in Beirut, Lebanon (1983). He may try to use information operations to undermine alliances and influence public opinion.

Tempo

One of the most powerful weapons available to the commander is speed. The unit that can consistently move and act faster than its enemy has a powerful advantage. The ability to plan, decide, execute, and assess faster than the enemy creates advantage that commanders can exploit.

In a military sense, there is more to speed than simply going fast, and there is a vital difference between acting rapidly and acting recklessly. Speed and time are closely related. In tactics, time is always of the utmost importance. Time that cannot be spent in action must be spent preparing for action and shaping conditions for decisive actions. If speed is a weapon, so is time. Speed and time create tempo. Tempo is the rate of military action and has significance only in relation to that of the enemy. When friendly tempo exceeds that of the enemy to react, friendly forces seize and maintain the initiative and have a marked advantage.

Tempo is not merely a matter of acting fastest or at the earliest opportunity. It is also a matter of timing—acting at the right time. The commander must be able to generate and maintain a fast pace when the situation calls for it and to recover when it will not hurt. Timing means knowing when to act and, equally important, when not to act. To be consistent, superiority in tempo must continue over time. It is not enough to move faster than the enemy only now and then. When the friendly force is not moving faster, the advantage and initiative passes to the enemy. Most forces can manage an intermittent burst of speed but must then halt for a considerable period to recover. During that halt, they are likely to lose their advantage. While a force cannot operate at full speed indefinitely, the challenge is to be consistently faster than the enemy.

To act consistently faster than the enemy, it is necessary to do more than move quickly. It is also necessary to make rapid transitions from one action to another. While there are many types of transitions in combat, the important thing to remember is that transitions produce friction. Reduction of friction minimizes the loss of tempo that is generated at the point of transition. A unit that can make transitions faster and more smoothly than another can be said to have greater relative speed.

Adapting

War is characterized by friction, uncertainty, disorder, and rapid change. Each situation is a unique combination of shifting factors that cannot be controlled with precision or certainty. A tactically proficient leader must be able to adapt actions to each situation. For adaptation to be effective, commanders must readily exploit the opportunities uncovered by subordinates. While making the best possible preparations, they must welcome and take advantage of unforeseen opportunities.

There are two basic ways to adapt. Sometimes the commander has enough situational awareness to understand a situation in advance and take preparatory action. This is anticipation. At other times he has to adapt to the situation on the spur of the moment without time for preparation. This is improvisation. A successful commander must be able to do both.

To anticipate, the commander must forecast future actions, at least to some extent. Forecasts are usually based on past experiences, learned through trial and error in training, exercises or actual combat. Planning is a form of anticipatory adaptation—adapting actions in advance. Another form of adaptation is immediate-action drills or standing operating procedures. These tools allow Marines to react immediately in a coordinated way to a broad variety of tactical situations. They provide the basis for adaptation.

Improvising is adjusting to a situation on the spur of the moment without any preparation. Improvisation requires creative, intelligent, and experienced leaders who have an intuitive appreciation for what will work and what will not. Improvisation is of critical importance to increasing speed. It requires commanders with a strong situational awareness and a firm understanding of their senior commander's intent so that they can adjust their own actions in accordance with the higher commander's desires.

Exploiting Success and Finishing

The successful commander exploits any advantage aggressively and relentlessly, not once, but repeatedly, until the opportunity arises for the finishing stroke. He uses this advantage to create opportunities. The commander builds on successfully exploited advantages to create new advantage, reflecting the changing situation resulting from both friendly and enemy adaptation. Advantages don't have to be large—small favorable circumstances exploited repeatedly and aggressively can quickly multiply into decisive advantages. In the same way, the commander exploits opportunities to create others. Victories are usually the result of aggressively exploiting some advantage or opportunity until the action becomes decisive.

Such victories are realized through development and maintenance of momentum and the successful attack of enemy critical vulnerabilities. Momentum is the increase of combat power, gained from seizing the initiative and attacking aggressively and rapidly. Once the commander decides to exploit an advantage, he makes every effort to build momentum until the offensive becomes overwhelming and the objective is achieved. He should not sacrifice momentum to preserve the alignment of advancing units and he should drive hard at gaps in the enemy defense. The commander should not waste time or combat power on enemy units that cannot jeopardize the overall mission, choosing instead to fix them with minimal forces and bypass them with his main force. The commander exploits enemy weaknesses such as tactical errors, faulty dispositions, assailable flanks, poor or no preparation, lack of support, numerical and equipment inferiority, low morale, and predictable operational patterns.

Subordinates must be accustomed through practice and training to seize opportunities on their own volition. For example, a major subordinate commander should have a complete understanding of the MAGTF commander's intent so that he will recognize a decisive opportunity and have the confidence to rapidly exploit the opportunity without further orders from the MAGTF commander. The commander exploits opportunities by conducting consolidation, exploitation or pursuit.

Exploiting advantages without applying the finishing stroke to defeat the enemy or achieve the objective cannot be decisive. Once the commander has created the opportunity to deliver the decisive blow he must strike the enemy vigorously and relentlessly, until the enemy is defeated. At the same time, the commander must exercise judgment to ensure that the force committed to the decisive action is not unduly or unintentionally made vulnerable to enemy counteraction. Rapid and accurate assessment by the MAGTF commander and his major subordinate commanders is critical in determining the appropriate time and place of the decisive action.

According to MCDP 1-3, tactical excellence is the hallmark of a Marine Corps leader. We fight and win in combat through our mastery of both the art and science of tactics. The art of tactics involves the creative and innovative use of maneuver warfare concepts, while the science of tactics requires skill in basic warfighting techniques and procedures. It is our responsibility as Marine leaders to work continuously to develop our own tactical proficiency and that of our Marines.

CHAPTER 7

The MAGTF in the Offense

Contents	
Purpose of Offensive Operations	**7-2**
Characteristics of Offensive Operations	**7-3**
Organization of the Battlespace	7-3
Organization of the Force	7-4
Types of Offensive Operations	**7-7**
Movement to Contact	7-7
Attack	7-10
Exploitation	7-14
Pursuit	7-15
Forms of Maneuver	**7-16**
Frontal Attack	7-17
Flanking Attack	7-18
Envelopment	7-19
Turning Movement	7-21
Infiltration	7-22
Penetration	7-23
Future Offensive Operations	**7-24**

> "Since I first joined the Marines, I have advocated aggressiveness in the field and constant offensive action. Hit quickly, hit hard and keep right on hitting. Give the enemy no rest, no opportunity to consolidate his forces and hit back at you. This is the shortest road to victory."
> —General H.M. "Howling Mad" Smith, USMC

The commander conducts offensive operations within the context of the single battle. The offense is the decisive form of warfare. While defensive operations can do great damage to an enemy, offensive operations are the means to a decisive victory. Offensive operations are conducted to take the initiative from the enemy, gain freedom of action, and mass effects to achieve objectives. These operations impose the commander's will on the enemy. Offensive operations allow the commander to impose his will on the enemy by shattering the enemy's moral, mental, and physical cohesion. The enemy loses his ability to fight as an effective, coordinated force as Marine Corps forces generate an overwhelming tempo by conducting a variety of rapid, focused, and unexpected offensive actions.

PURPOSE OF OFFENSIVE OPERATIONS

The offense is undertaken to gain, maintain, and exploit the initiative—thus causing the enemy to react to our actions. The focus of offensive operations is the enemy, not seizure of terrain. Even in the defense, a commander must take every opportunity to seize the initiative by offensive action. Offensive operations are conducted to—

- Destroy enemy forces and equipment.
- Deceive and divert the enemy.
- Deprive the enemy of resources.
- Gain information.
- Fix the enemy in place.
- Seize key terrain.
- Produce a reaction from the enemy.
- Disrupt enemy actions or preparations.

Successful offensive operations—

- Avoid the enemy's strength and attack his weakness by massing combat power or its effects against the enemy's critical vulnerabilities.
- Isolate the enemy from his sources of support—both moral and physical—to include logistics, fires, command and control, and reinforcements.
- Strike the enemy from an unexpected direction, disrupting his plan.
- Aggressively exploit every advantage.
- Overwhelm the enemy commander's ability to observe, orient, decide, and act.
- Employ accurate and timely assessment of effects against the enemy to exploit success.

The Marine Corps' warfighting philosophy is offensive in nature, focuses on the threat, uses speed to seize the initiative, and surprise to degrade the enemy's ability to resist. Offensive operations require the attacker to weight the main effort with superior combat power. The requirement to concentrate and the need to have sufficient forces available to exploit success imply accepting risk elsewhere. Local superiority must be created by maneuver, speed, surprise, and economy of force.

Before conducting offensive operations, the commander seeks to discover where the enemy is most vulnerable through reconnaissance and surveillance. Shaping actions set the conditions for decisive action by disrupting the enemy's command

and control, limiting his ability to apply combat power, and further exposing weaknesses in the defense. Shaping actions should place the enemy at the greatest disadvantage possible. The commander directs the battle from a position that allows him to develop a firsthand impression of the course of the battle. He provides personal leadership and inspires confidence at key points in the battle.

The fundamentals of offensive action are general rules evolved from logical and time-proven application of the principles of war to the offense. Many of the fundamentals are related and reinforce one another as follows:

- Orient on the enemy.
- Gain and maintain contact.
- Develop the situation.
- Concentrate superior firepower at the decisive time and place.
- Achieve surprise.
- Exploit known enemy weaknesses.
- Seize or control key terrain.
- Gain and retain the initiative.
- Neutralize the enemy's ability to react.
- Advance by fire and maneuver.
- Maintain momentum.
- Act quickly.
- Exploit success.
- Be flexible.
- Be aggressive.
- Provide for the security of the force.

CHARACTERISTICS OF OFFENSIVE OPERATIONS

Organization of the Battlespace

Deep operations are conducted using maneuver forces, fires, and information operations. They seek to create windows of opportunity for decisive maneuver and are designed to restrict the enemy's freedom of action, disrupt the coherence and tempo of his operations, nullify his firepower, disrupt his command and control, interdict his supplies, isolate or destroy his main forces, and break his morale.

The enemy is most easily defeated by fighting him close and deep simultaneously while protecting the MAGTF rear area. Well-orchestrated deep operations, integrated with simultaneous close operations, may be executed with the goal of defeating the enemy outright or setting the conditions for successful future close operations. Deep operations enable friendly forces to choose the time, place, and method for close operations. Deep operations may include—

- Deception.
- Deep interdiction through deep fires, deep maneuver, and deep air support.
- Deep surveillance and target acquisition.
- Information operations.
- Offensive antiair warfare.

Close operations are required for decisive and lasting effects on the battlefield. The MAGTF commander shapes the course of the battle and takes decisive action, deciding when and where to commit the main effort to achieve mission success. The MAGTF commander picks a combination of the types of offensive operations and forms of maneuver to use at the critical time and place to defeat the enemy. Commanders weight their combination of options to mass effects. For example, commanders may fix a part of the enemy force with a frontal attack by a smaller combined arms force while maneuvering the rest of the force in an envelopment to defeat the enemy force. The reserve enters the action offensively at the proper place and moment to exploit success. The reserve provides the source of additional combat power to commit at the decisive moment.

Rear area operations protect assets in the rear area to support the force. Rear area operations encompass more than just rear area security. While rear area operations provide security for personnel, materiel, and facilities in the rear area, their sole purpose is to provide uninterrupted support to the force as a whole. Rear area operations enhance a force's freedom of action while it is involved in the close and deep fight and extend the force's operational reach. The primary focus of rear area operations during the offensive is to maintain momentum and prevent the force from reaching a culminating point.

Organization of the Force

The commander will normally organize his force differently depending on the type of offensive operation he is conducting. There are four basic types of forces: security forces, main body, reserve, and sustainment forces.

The commander may use security forces to—

- Gain and maintain enemy contact.
- Protect the main battle force's movement.
- Develop the situation before committing the main battle force.

Security forces are assigned cover, guard or screen missions. Operations of security forces must be an integral part of the overall offensive plan. The element of the MAGTF assigned as the security forces depends on the factors of METT-T. Security forces are discussed in detail in chapter 11.

The main body constitutes the bulk of the commander's combat power. It is prepared to respond to enemy contact with the security forces. Combat power that can be concentrated most quickly, such as fires, is brought to bear while maneuver units move into position. The main body maintains an offensive spirit throughout the battle, looking to exploit any advantageous situations. The main body engages the enemy as early as possible unless fires are withheld to prevent the loss of surprise. Commanders make maximum use of fires to destroy and disrupt enemy formations. As the forces close, the enemy is subjected to an ever-increasing volume of fires from the main body and all supporting arms.

The commander uses his reserve to restore momentum to a stalled attack, defeat enemy counterattacks, and exploit success. The reserve provides the commander the flexibility to react to unforeseen circumstances. Once committed, the reserve's actions normally become the decisive operation, and every effort is made to reconstitute another reserve from units made available by the revised situation.

In the attack, the combat power allocated to the reserve depends primarily on the level of uncertainty about the enemy, especially the strength of any expected enemy counterattacks. The commander only needs to resource a small reserve to respond to unanticipated enemy reactions when he has detailed information about the enemy. When the situation is relatively clear and enemy capabilities are limited, the reserve may consist of a small fraction of the command. When the situation is vague, the reserve may initially contain the majority of the commander's combat power.

In an attack, the commander generally locates his reserve to the rear of the main effort. However, it must be able to move quickly to areas where it is needed in different contingencies. This is most likely to occur if the enemy has strong counterattack forces. For heavy reserve forces, the key factor is cross-country mobility or road networks. For light forces, the key factor is the road network, if trucks are available, or the availability of landing zones for helicopterborne forces. The commander prioritizes the positioning of his reserve to counter the worst case enemy counterattack first, then to reinforce the success of the decisive operation.

The commander task-organizes his sustainment forces to the mission. He decentralizes the execution of sustainment support, but that support must be continuously available to the main body. This includes using preplanned logistics packages. Aerial resupply may also be necessary to support large-scale movements to contact or to maintain the main body's momentum. Combat trains containing fuel, ammunition, medical, and maintenance assets move with their parent unit. Fuel and ammunition stocks remain loaded on tactical vehicles in the combat trains so they can instantly move. Aviation assets may use forward operating bases (including forward arming and refueling points and rapid ground refueling sites) to reduce aircraft turnaround time. The commander will frequently find that his main supply routes become extended as the operation proceeds.

In an attack, the commander tries to position his sustainment forces well forward. From these forward locations they can sustain the attacking force, providing priority of support to the decisive operation. As the attacking force advances, sustainment forces displace forward as required to shorten the supply lines to ensure uninterrupted support to maneuver units. The size of the force a commander devotes to sustainment force security depends on the threat. A significant enemy threat requires the commander to provide a tactical combat force.

During periods of rapid movement sustainment forces may be attached to the moving or attacking force. Alternatively, sustainment forces may follow the moving or attacking force in an echeloned manner along main supply routes. Transportation and supplies to sustain the moving or attacking force become increasingly important as the operation progresses. As supply lines lengthen, the condition of lines of communications and the conduct of route and convoy security can become problems. The largest possible stocks of fuel, spare parts, and ammunition should accompany the moving or attacking force so that it does not lose momentum because of a lack of support. The offensive operation may be limited more by vehicle mechanical failures and the need for fuel than by combat losses or a lack of ammunition. Therefore, direct support maintenance support teams accompany the moving or attacking force to repair disabled vehicles or evacuate them to maintenance collection points for repair by general support maintenance units. The commander may also use helicopters to move critical supplies forward.

TYPES OF OFFENSIVE OPERATIONS

There are four types of offensive operations—movement to contact, attack, exploitation, and pursuit. These operations may occur in sequence, simultaneously or independently across the depth of the battlespace. For example, a movement to contact may be so successful that it immediately leads to an exploitation or an attack may lead directly to pursuit. See figure 7-1.

These types of offensive operations are rarely all performed in one campaign or in the sequence presented in this chapter. Nor are the dividing lines between the types of offensive operations as distinct in reality as they are in a doctrinal publication. The successful commander uses the appropriate type of offensive operation for his mission and situation, not hesitating to change to another type if the battle dictates. The goal is to move to exploitation and pursuit as rapidly as possible. The commander seeks to take advantage of enemy weaknesses and maneuver to a position of advantage, creating the conditions that lead to exploitation.

Movement to Contact

Movement to contact seeks to gain or regain contact with the enemy and develop the situation. Movement to contact helps the commander to understand the battlespace. It allows him to make initial contact with the enemy with minimum forces, thereby avoiding an extensive engagement or battle before he is prepared for decisive action. When successfully executed, it allows the commander to strike the enemy at the time and place of his choosing. A movement to contact ends when the commander has to deploy the main body—to conduct an attack or establish a defense.

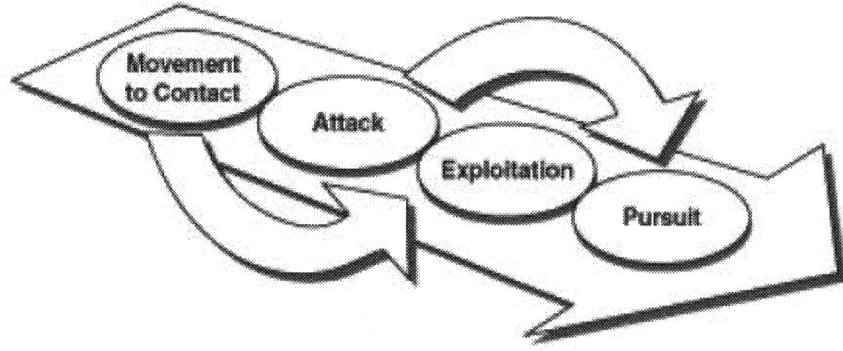

Figure 7-1. Types of Offensive Operations.

A force conducting movement to contact normally organizes in an approach march formation, with advance, flank, and rear security elements protecting the main body. See figure 7-2. The main body contains the bulk of the MAGTF's forces. The advance force, flank, and rear security formations may consist of aviation or ground combat units (one or both as individual elements or as task-organized combined arms teams) and appropriate combat service support organizations, based on the factors of METT-T.

As the purpose of movement to contact is to gain contact with the enemy, the MAGTF commander will normally designate the advance force as the main effort. As contact with the enemy is made and the situation develops, the MAGTF commander has two options. If he decides that this is not the time or place to offer battle, he bypasses the enemy and the advance force remains the main effort. When bypassing an enemy unit it may be necessary to task a subordinate unit to fix or block the bypassed enemy. As the second option, if the MAGTF commander determines that his shaping actions have set the conditions for decisive action, he will shift the main effort—probably to the main body or a unit in the main body. During movement to contact, the MAGTF commander may designate all or part of the main body as the MAGTF reserve.

The MAGTF commander may use the ACE to exploit the situation as it is developed by the advance force. Aviation forces can attack enemy forces involved in a meeting engagement or fix enemy forces while the advance force makes contact, allowing the main body to maneuver without becoming decisively

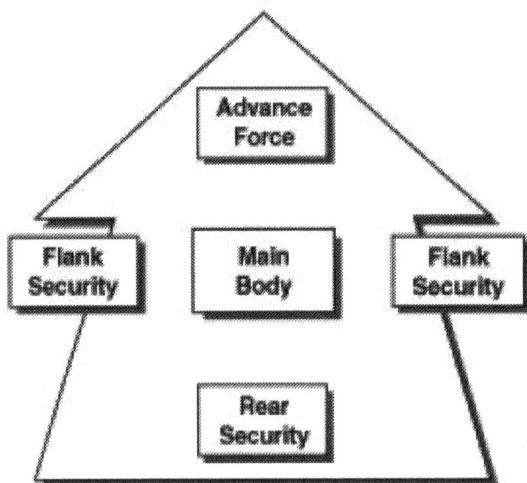

Figure 7-2. Movement to Contact.

engaged. Aviation forces can also attack second echelon forces, limiting the enemy's ability to reinforce his first echelon or fight in depth.

Even if the MAGTF commander properly develops the situation during a movement to contact, he may be faced with a meeting engagement. Meeting engagements are clashes that take place at unexpected places and times when forces are not fully prepared for battle. A meeting engagement may result in confusion, delay or even in the premature employment of the main body before the MAGTF commander has set the conditions for decisive action. The premature employment of the main body slows the MAGTF's tempo of operations and may cause it to lose the initiative.

When organizing his forces for movement to contact, the MAGTF commander considers span of control, communications, and the capabilities of the major subordinate commands. Both the GCE and the ACE can provide forces for all the formations. The GCE can gain and maintain physical contact with the enemy. The ACE can establish initial visual and electromagnetic contact with the enemy at extended ranges.

The MAGTF commander should give the major subordinate commanders the widest latitude to conduct movement to contact. This allows them the freedom to maneuver and exercise initiative in developing the situation once contact with the enemy is made. Unnecessary subdivision of the MAGTF's battlespace or constraints placed on the major subordinate commands for fire and maneuver within their respective AO may only slow the MAGTF's tempo. The MAGTF commander must consider airspace requirements and the capability of the Marine air command and control system during movement to contact.

The frontage assigned to the unit conducting a movement to contact must allow it sufficient room to deploy its main body, but not be so wide that large enemy forces might be inadvertently bypassed. The unit's frontage will be affected by the MAGTF commander's guidance on bypassing enemy forces or the requirement to clear his zone of all enemy forces. The MAGTF commander must ensure that the desired rate of advance is supportable by the CSSE to ensure that he does not reach a culminating point or an unplanned operational pause.

The MAGTF may encounter enemy forces too small to threaten the main body and that do not require its deployment with the resulting loss of momentum. While such enemy forces may pose little threat against the ground or ACE's combat power, they may pose a serious threat to the CSSE. The MAGTF commander must establish criteria on the size, nature or type of enemy activity that the main body may bypass and how assigned follow and support forces will deal with bypassed enemy forces.

The MAGTF commander identifies potential danger areas where his forces may make contact with the enemy, such as likely enemy defensive locations, engagement areas, observation posts, and natural and artificial obstacles. The MAGTF's reconnaissance and surveillance plan must provide for coverage of these danger areas. The MAGTF commander must recognize the enemy's most dangerous course of action and be prepared to focus his combat power at those times and places where the MAGTF is most vulnerable.

Because the MAGTF may be engaged by the enemy's air forces well before coming into contact with the enemy's ground forces, offensive air support (close air support and air interdiction specifically) and antiair warfare functions are essential to success of the movement to contact. The ACE must gain and maintain contact with the enemy. The MAGTF commander may also task the ACE to conduct deep reconnaissance to determine or confirm the location of enemy reserve, follow-on, and support forces. If he has enough intelligence to target these enemy forces, he may task the ACE to conduct an attack.

The CSSE must provide the full range of combat service support during movement to contact without slowing or jeopardizing the MAGTF's tempo. The support must be responsive and provide only what is needed—it should not encumber movement or maneuver. This requires a fine balance between push- and pull-logistics. The CSSE must take advantage of natural pauses during the course of the MAGTF's movement to replenish expended resources. Support for security forces is more difficult due to their separation from the main body and the need for self-contained maintenance and supply capabilities. This may also require a greater reliance on ACE assets to ensure prompt and effective support.

The MAGTF commander must ensure that once contact with the enemy has been gained, that it is maintained with whatever assets available. Aviation and ground combat assets may have to shift to maintain contact. The CSSE can also assist by ensuring that combat units do not reach a culminating point.

As the situation becomes clearer, the MAGTF commander is better able to determine how best to exploit the opportunity. Ideally, this means shifting to an attack, exploitation or pursuit. The major subordinate commanders must be ready to support a transition from one type of offensive operation to another.

Attack

An attack is an offensive operation characterized by coordinated movement, supported by fire, conducted to defeat, destroy, neutralize or capture the enemy. An attack may be conducted to seize or secure terrain. Focusing combat power against the enemy with a tempo and intensity that the enemy cannot match, the commander attacks to shatter his opponent's will, disrupt his cohesion, and to gain the initiative. If an attack is successful, the enemy is no longer capable of—or willing to offer—meaningful resistance.

Attacks rarely develop exactly as planned. As long as the enemy has any freedom of action, unexpected difficulties will occur. As the attack progresses, control must become increasingly decentralized to subordinate commanders to permit them to meet the rapidly shifting situation. This is achieved through the use of the commander's intent and mission tactics. The commander sets conditions for a successful attack by attacking enemy fire support assets, command and control assets and support facilities, and front-line units. These fires protect the main effort and restrict the enemy's ability to counterattack. During the final stages of the attack, the main effort may rely primarily on organic fires to overcome remaining enemy resistance. The attack culminates in a powerful and violent assault. The commander immediately exploits his success by continuing the attack into the depth of the enemy defense to disrupt his cohesion.

Attacks can be hasty or deliberate based on the degree of preparation, planning, and coordination involved prior to execution. The distinction between hasty and deliberate attacks is a relative one.

A hasty attack is an attack when the commander decides to trade preparation time for speed to exploit an opportunity. A hasty attack takes advantage of audacity, surprise, and speed to achieve the commander's objectives before the enemy can effectively respond. The commander launches a hasty attack with the forces at hand or in immediate contact with the enemy and with little preparation before the enemy can concentrate forces or prepare an effective defense.

By necessity, hasty attacks do not employ complicated schemes of maneuver and require a minimum of coordination. Habitual support relationships, standing operating procedures, and battle drills contribute to increased tempo and the likelihood of success of the hasty attack. Unnecessary changing of the task organization of the force should be avoided to maintain momentum.

A deliberate attack is a type of offensive action characterized by pre-planned and coordinated employment of firepower and maneuver to close with and destroy the enemy. Deliberate attacks usually include the coordinated use of all available resources. Deliberate attacks are used when the enemy cannot be defeated with a hasty attack or there is no readily apparent advantage that must be rapidly exploited.

Main and supporting efforts and the forward positioning of resources are planned and coordinated throughout the battlespace to ensure the optimal application of the force's combat power. The commander must position follow-on forces and the reserve to best sustain the momentum of the attack. Deliberate attacks may include time for rehearsals and refinement of attack plans. The commander must weigh the advantages of a deliberate attack with respect to the enemy's ability to create or improve his defenses, develop his intelligence picture or take counteraction.

Commanders conduct various types of attack to achieve different effects. A single attack that results in the complete destruction or defeat of the enemy is rare. The commander must capitalize on the resulting disruption of the enemy's defenses through exploitation to reap the benefits of a successful attack.

Spoiling Attack

A spoiling attack is a tactical maneuver employed to seriously impair a hostile attack while the enemy is in the process of forming or assembling for an attack. A spoiling attack is usually an offensive action conducted in the defense. See chapter 8.

Counterattack

A counterattack is a limited-objective attack conducted by part or all of a defending force to prevent the enemy from attaining the objectives of his attack. It may be conducted to regain lost ground, destroy enemy advance units, and wrest the initiative from the enemy. It may be the precursor to resuming offensive operations. See chapter 8.

Feint

A feint is a limited-objective attack made at a place other than that of the main effort with the aim of distracting the enemy's attention away from the main effort. A feint is a supporting attack that involves contact with the enemy. A feint must be sufficiently strong to confuse the enemy as to the location of the main attack. Ideally, a feint causes the enemy to commit forces to the diversion and away from the main effort. A unit conducting a feint usually attacks on a wider front than normal with a consequent reduction in mass and depth. A unit conducting a feint normally keeps only a minimal reserve to deal with unexpected developments.

Demonstration

A demonstration is an attack or a show of force on a front where a decision is not sought. Its aim is to deceive the enemy. A demonstration, like a feint, is a supporting attack. A demonstration, unlike a feint, does not make contact with the enemy. The commander executes a demonstration by an actual or simulated massing of combat power, troop movements or some other activity designed to indicate the preparations for or beginning of an attack at a point other than the main effort. Demonstrations are used frequently in amphibious operations to draw enemy forces away from the actual landing beaches or to fix them in place. Demonstrations and feints increase the enemy's confusion while conserving combat power for the main and supporting efforts.

Reconnaissance in Force

A reconnaissance in force is a deliberate attack made to obtain information and to locate and test enemy dispositions, strengths, and reactions. It is used when knowledge of the enemy is vague and there is insufficient time or resources to develop the situation. While the primary purpose of a reconnaissance in force is to gain information, the commander must be prepared to exploit opportunity. Reconnaissance in force usually develops information more rapidly and in more detail than other reconnaissance methods. If the commander must develop the enemy situation along a broad front, reconnaissance in force may consist of strong probing actions to determine the enemy situation at selected points.

The commander may conduct reconnaissance in force as a means of keeping pressure on the defender by seizing key terrain and uncovering enemy weaknesses. The reconnoitering force must be of a size and strength to cause the enemy to react strongly enough to disclose his locations, dispositions, strength, planned fires, and planned use of the reserve. Since a reconnaissance in force is conducted when knowledge of the enemy is vague, a task-organized combined arms force normally is used. Deciding whether to reconnoiter in force, the commander considers—

- His present information on the enemy and the importance of additional information.
- Efficiency and speed of other intelligence collection assets.
- The extent his future plans may be divulged by the reconnaissance in force.
- The possibility that the reconnaissance in force may lead to a decisive engagement that the commander does not desire.

Raid

A raid is an attack, usually small scale, involving a penetration of hostile territory for a specific purpose other than seizing and holding terrain. It ends with the planned withdrawal upon completion of the assigned mission. The organization and composition of the raid force are tailored to the mission. Raids are characterized by surprise and swift, precise, and bold action. Raids are typically conducted to—

- Destroy enemy installations and facilities.
- Disrupt enemy command and control or support activities.
- Divert enemy attention.
- Secure information.

Raids may be conducted in the defense as spoiling attacks to disrupt the enemy's preparations for attack; during delaying operations to further delay or disrupt the enemy or with other offensive operations to confuse the enemy, divert his attention or disrupt his operations. Raids require detailed planning, preparation, and special training.

Exploitation

Exploitation is an offensive operation that usually follows a successful attack and is designed to disorganize the enemy in depth. The exploitation extends the initial success of the attack by preventing the enemy from disengaging, withdrawing, and reestablishing an effective defense. The exploitation force expands enemy destruction through unrelenting pressure thus weakening his will to resist. The exploitation is characterized by initiative, boldness, and the unhesitating employment of uncommitted forces.

The commander must prepare to exploit the success of every attack without delay. In the hasty attack, the force in contact normally continues the attack, transitioning to exploitation. In the deliberate attack, the commander's principal tool for the exploitation is normally the reserve. At the MAGTF level, aviation forces may support the reserve or be additionally tasked as the exploitation force. The commander retains only those reserves necessary to ensure his flexibility of operation, continued momentum in the advance, and likely enemy responses to the exploitation. The reserve is generally positioned where it can exploit the success of the main effort or supporting efforts. Exploitation forces execute bold, aggressive, and rapid operations using the commander's intent and mission tactics.

The decision to commence the exploitation requires considerable judgment, intuition, and situational awareness by the commander. Committing the exploitation force prematurely or too late may fail to exploit the opportunity presented by a successful attack. Conditions favorable for an exploitation may include—

- Increased number of enemy prisoners of war.
- Absence of organized defenses.

- Absence of accurate enemy massed direct and indirect fires.
- Loss of enemy cohesion upon contact.
- Capture, desertion or absence of enemy commanders and senior staff officers.

Typical objectives for the exploitation force include command posts, reserves, seizure of key terrain, and the destruction of combat support and service support units deep in the enemy's rear. The destruction or defeat of these objectives further disrupt and disorganize the enemy, preventing reconstitution of the defense or the enemy's force. The commander must be prepared to assess the effects of his exploitation and determine when the time is at hand to commence the pursuit of the enemy.

Pursuit

A pursuit is an offensive operation designed to catch or cut off a hostile force attempting to escape, with the aim of destroying it. Pursuits often develop from successful exploitation operations when the enemy defenses begin to disintegrate. A pursuit may also be initiated when the enemy has lost his ability to fight effectively and attempts to withdraw.

Since the conditions that allow for pursuit can seldom be predicted, a pursuit force is not normally established ahead of time. The commander must quickly designate appropriate forces to conduct and support pursuit operations or the exploitation force may continue as the pursuit force. A pursuit is normally made up of a direct pressure force and an encircling force. See figure 7-3 on page 7-16.

These forces are similar to a hammer and anvil. The direct pressure force is like the hammer. It is usually a powerful maneuver force that maintains continuous contact with the retreating enemy, driving the enemy before them. The encircling force serves as the anvil. The encircling force requires sufficient mobility and speed to get itself into position ahead of or on the flank of the fleeing enemy to halt and fix the enemy in place. Aviation forces are particularly well-suited to act as the encircling force. By using its superior tactical mobility and agility in concert with its potent firepower, aviation forces can destroy enemy forces, interdict lines of retreat, and add to the demoralization of the enemy force.

Pursuits are pushed to the utmost limits of endurance of troops, equipment, and supplies. If the pursuit force is required to pause for any reason, the enemy has an opportunity to break contact, reorganize, and establish organized defenses. Pursuit, like exploitation, must be conducted relentlessly. Highly mobile and versatile combat service support forces are particularly critical to sustaining a relentless pursuit and preventing the MAGTF from reaching its culminating point before the enemy is completely defeated.

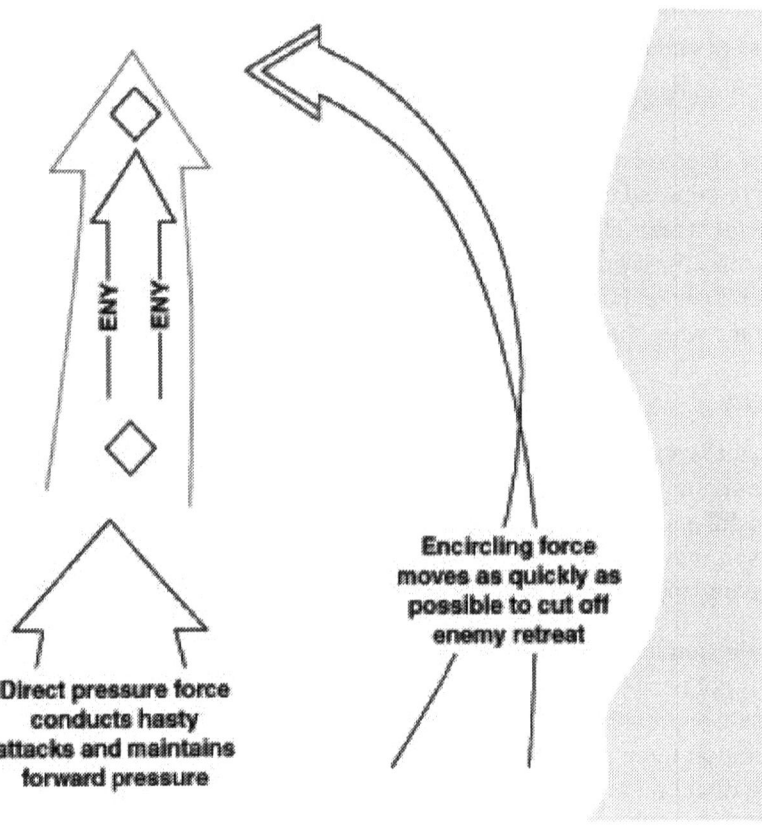

Figure 7-3. Pursuit.

Forms of Maneuver

The forms of offensive maneuver are the basic techniques a force conducting offensive operations uses to gain advantage over the enemy. Each form of maneuver has a resultant effect on the enemy. The MAGTF commander chooses the form of maneuver that fully exploits all the dimensions of the battlespace and best accomplishes his mission. He generally chooses one of these as a foundation upon which to build a course of action.

The MAGTF commander organizes and employs the ACE, GCE or CSSE to best support the chosen form of maneuver. The GCE and ACE are the two combat arms of the MAGTF. They execute tactical actions to support or accomplish the MAGTF commander's mission. Either can be used as a maneuver force or a source of fires as the MAGTF commander applies combined arms. The MAGTF commander may task-organize aviation and ground combat units, along with combat service support units, under a single commander to execute the form of offensive maneuver selected. Aviation forces may be comprised of fixed-wing aircraft, rotary-wing aircraft or a combination with GCE and CSSE units attached or in support.

Frontal Attack

A frontal attack is an offensive maneuver where the main action is directed against the front of the enemy forces. It is used to rapidly overrun or destroy a weak enemy force or fix a significant portion of a larger enemy force in place over a broad front to support a flanking attack or envelopment. It is generally the least preferred form of maneuver because it strikes the enemy where he is the strongest. See figure 7-4. It is normally used when commanders possess overwhelming combat power and the enemy is at a clear disadvantage.

For deliberate attacks, the frontal attack may be the most costly form of maneuver since it exposes the attacker to the concentrated fires of the defender while limiting the effectiveness of the attacker's own fires. As the most direct form of maneuver, however, the frontal attack is useful for overwhelming light defenses, covering forces or disorganized enemy forces.

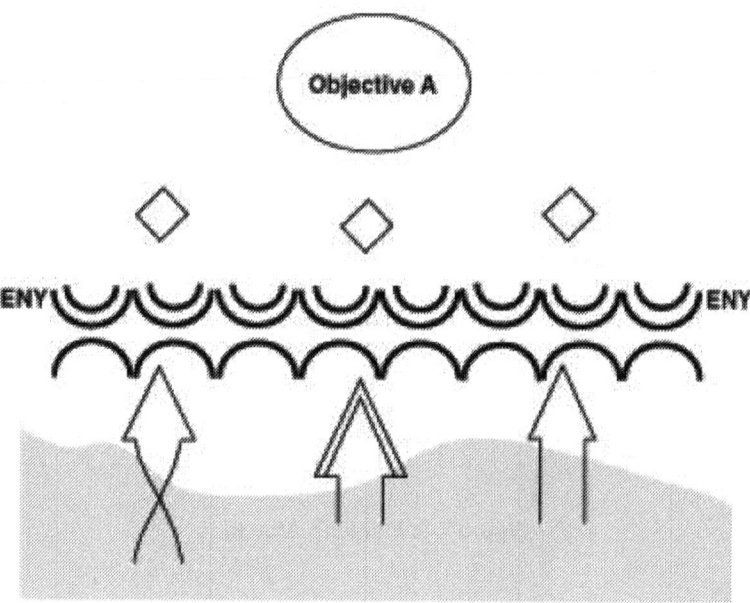

Figure 7-4. Frontal Attack.

Frontal attacks may be used by supporting efforts to fix the enemy in place and enable the main effort to maneuver to a position of advantage during an envelopment or a flanking attack. A frontal attack can create a gap through which the attacking force can conduct a penetration. Frontal attacks are often used with feints and demonstrations. Aviation forces and supporting arms are often used to create gaps with fires in the enemy's front or to prevent or delay enemy reinforcements reaching the front lines.

Flanking Attack

A flanking attack is a form of offensive maneuver directed at the flank of an enemy force. See figure 7-5. A flank is the right or left side of a military formation and is not oriented toward the enemy. It is usually not as strong in terms of forces or fires as is the front of a military formation. A flank may be created by the attacker through the use of fires or by a successful penetration. It is similar to an envelopment but generally conducted on a shallower axis. Such an attack is designed to defeat the enemy force while minimizing the effect of the enemy's frontally oriented combat power. Flanking attacks are normally conducted with the main effort directed at the flank of the enemy. Usually, there is a supporting effort that engages by fire and maneuver the enemy force's front while the main effort maneuvers to attack the enemy's flank. This supporting effort diverts the enemy's attention from the threatened flank. It is often used for a hasty attack or meeting engagement where speed and simplicity are paramount to maintaining battle tempo and, ultimately, the initiative.

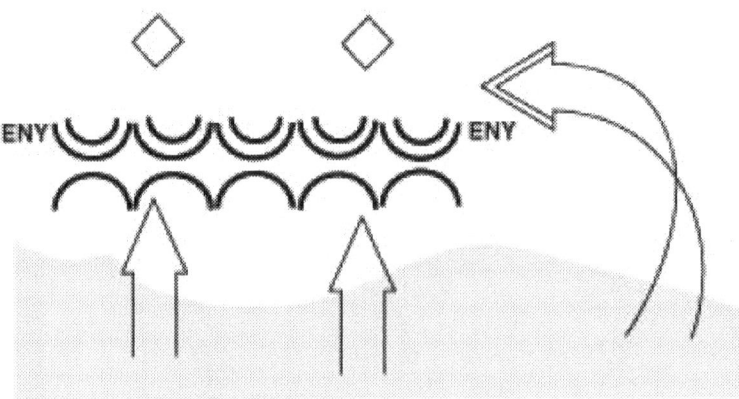

Figure 7-5. Flanking Attack.

Envelopment

An envelopment is a form of offensive maneuver by which the attacker bypasses the enemy's principal defensive positions to secure objectives to the enemy's rear. See figure 7-6 and figure 7-7 on page 7-20. The enemy's defensive positions may be bypassed using ground, waterborne or vertical envelopment. An envelopment compels the defender to fight on the ground of the attacker's choosing. It requires surprise and superior mobility relative to the enemy. The operational reach and speed of aviation forces, coupled with their ability to rapidly mass effects on the enemy, make them an ideal force to conduct an envelopment. An envelopment is designed to—

- Strike the enemy where he is weakest (critical vulnerabilities).
- Strike the enemy at an unexpected place.
- Attack the enemy rear.
- Avoid the enemy's strengths.
- Disrupt the enemy's command and control.
- Disrupt the enemy's logistics effort.
- Destroy or disrupt the enemy's fire support assets.
- Sever the enemy's lines of communications.
- Minimize friendly casualties.

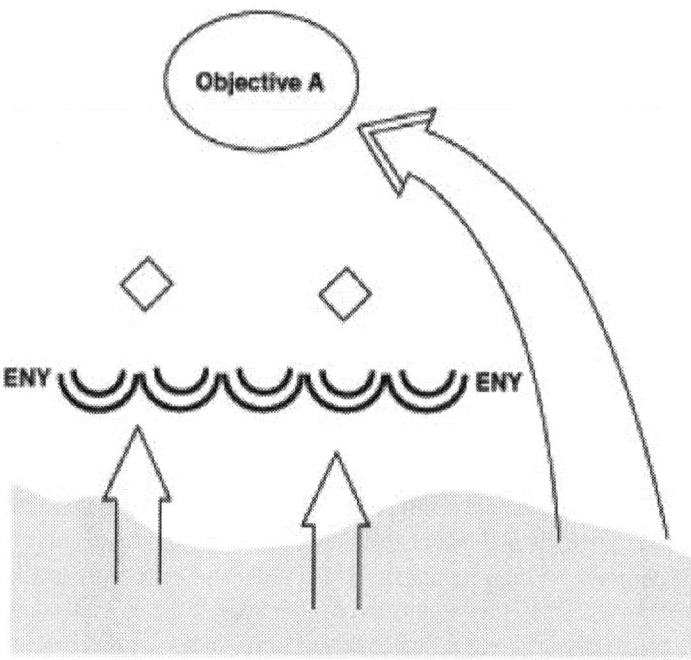

Figure 7-6. Single Envelopment.

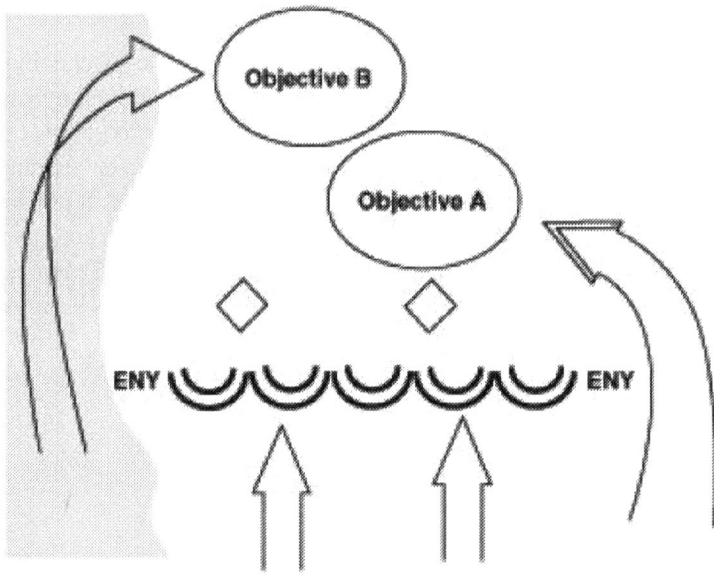

Figure 7-7. Double Envelopment.

When enveloping, the commander applies strength against weakness by maneuvering the main effort around or over an enemy's main defenses. Envelopments normally require a supporting effort to fix the enemy, prevent his escape, and reduce his ability to react against the main effort. They fix the enemy by forcing him to fight in multiple directions simultaneously or by deceiving him regarding the location, timing or existence of the main effort. Supporting efforts must be of sufficient strength to ensure these tasks are successful, as the success of the attack often depends on the effects achieved by the supporting effort.

An envelopment is conducted at sufficient depth so that the enemy does not have time to reorient his defenses before the commander concentrates his force for the attack on the objective. Because of their ability to rapidly mass, aviation forces are particularly well-suited to function as the enveloping force or to enable the success of the enveloping force.

The commander may choose to conduct a double envelopment. Double envelopments are designed to force the enemy to fight in two or more directions simultaneously to meet the converging axis of the attack. It may lead to the encirclement of the enemy force so the commander must be prepared to contain and defeat any breakout attempts. The commander selects multiple objectives to the rear of the enemy's defense and the enveloping forces use different routes to attack, seize or secure those objectives.

Turning Movement

A turning movement is a form of offensive maneuver where the attacker passes around or over the enemy's principal defensive positions to secure objectives deep in the enemy's rear. See figure 7-8. Normally, the main effort executes the turning movement as the supporting effort fixes the enemy in position. A turning movement differs from an envelopment in that the turning force usually operates at such distances from the fixing force that mutual support is unlikely. The turning force must be able to operate independently.

The goal of a turning movement is to force the enemy to abandon his position or reposition major forces to meet the threat. Once "turned" the enemy loses his advantage of fighting from prepared positions on ground of his choosing. Typical objectives of the main effort in a turning movement may include—

- Critical logistic sites.
- Command and control nodes.
- Lines of communications.

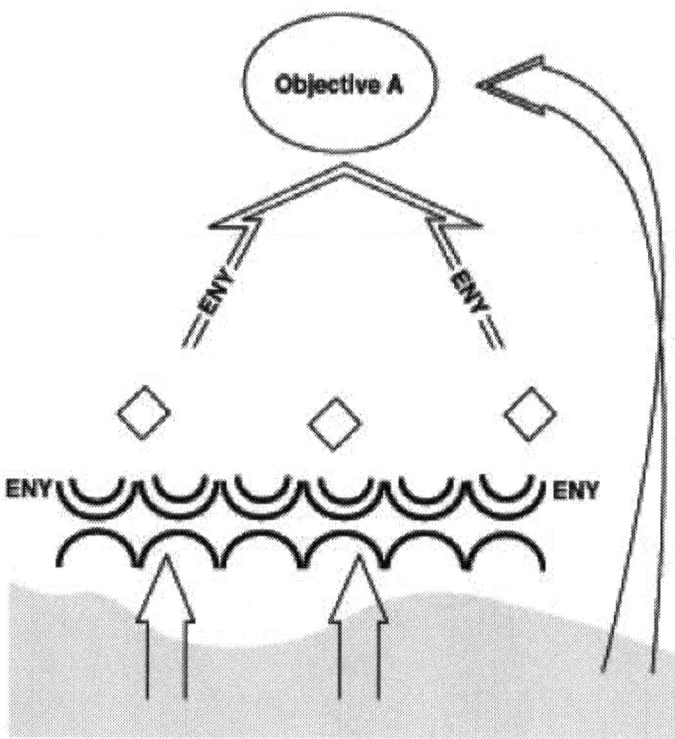

Figure 7-8. Turning Movement.

Using operational maneuver from the sea, the MAGTF is particularly well-suited to conduct a turning movement for the joint force commander. The ACE's speed and agility allow it to mass at the necessary operational depth to support the MAGTF commander's plan.

Infiltration

Infiltration is a form of maneuver where forces move covertly through or into an enemy area to attack positions in the enemy's rear. This movement is made, either by small groups or by individuals, at extended or irregular intervals. Forces move over, through or around enemy positions without detection to assume a position of advantage over the enemy. See figure 7-9. Infiltration is normally conducted with other forms of maneuver. The commander orders an infiltration to move all or part of his force through gaps in the enemy's defense to—

- Achieve surprise.
- Attack enemy positions from the flank or rear.
- Occupy a position from which to support the main attack by fire.
- Secure key terrain.
- Conduct ambushes and raids in the enemy's rear area to harass and disrupt his command and control and support activities.
- Cut off enemy forward units.

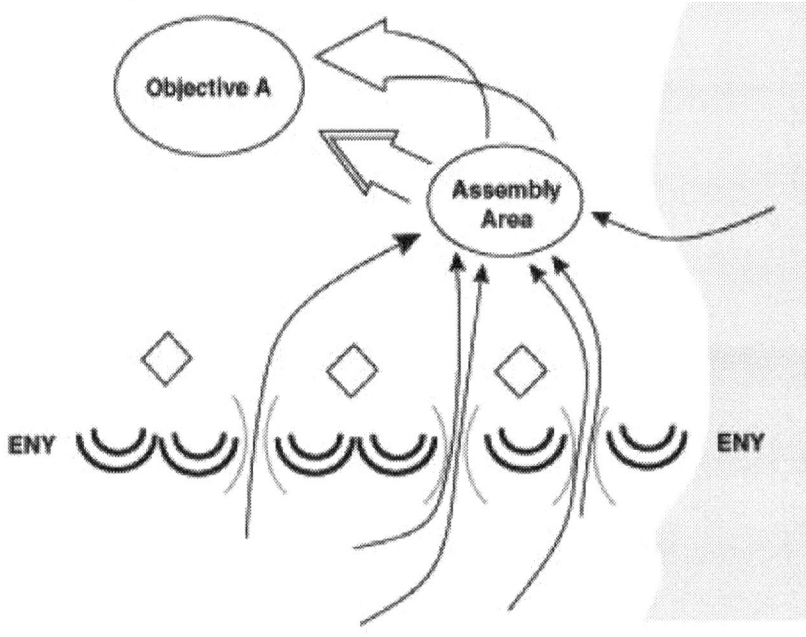

Figure 7-9. Infiltration.

Infiltrations normally take advantage of limited visibility, rough terrain or unoccupied or unobserved areas. These conditions often allow undetected movement of small elements when the movement of the entire force would present greater risks. The commander may elect to conduct a demonstration, feint or some other form of deception to divert the enemy's attention from the area to be infiltrated.

To increase control, speed, and the ability to mass combat power, a force infiltrates by the largest possible units compatible with the need for stealth, enemy capabilities and speed. Infiltrating forces may depend heavily on aviation forces for aerial resupply and close air support.

The infiltrating force may be required to conduct a linkup or series of linkups after infiltrating to assemble for its subsequent mission. Infiltration requires extremely detailed and accurate information about terrain and enemy dispositions and activities. The plan for infiltration must be simple, clear, and carefully coordinated.

Penetration

A penetration is a form of offensive maneuver where an attacking force seeks to rupture the enemy's defense on a narrow front to disrupt the defensive system. Penetrations are used when enemy flanks are not assailable or time, terrain or the enemy's disposition does not permit the employment of another form of maneuver. Successful penetrations create assailable flanks and provide access to the enemy's rear. A penetration generally occurs in three stages:

- Rupturing the position.
- Widening the gap.
- Seizing the objective.

A penetration is accomplished by concentrating overwhelmingly superior combat power on a narrow front and in depth. As the attacking force ruptures the enemy's defenses, units must be tasked to secure the shoulders of the breach and ultimately widening the gap for follow-on units. Rupturing the enemy position and widening the gap are not in themselves decisive. The attacker must exploit the rupture by attacking into the enemy's rear or attacking laterally to roll up the enemy's positions. See figure 7-10 on page 7-24. The shock action and mobility of a mechanized force and aviation forces are useful in rupturing the enemy's position and exploiting that rupture.

The commander may conduct multiple penetrations. Exploitation forces may converge on a single, deep objective or seize independent objectives. When it is

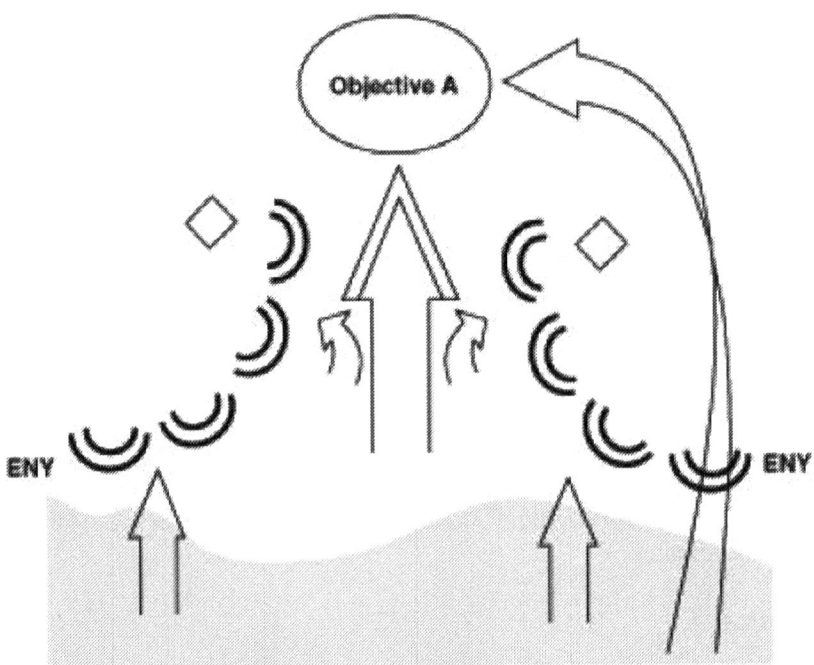

Figure 7-10. Penetration.

impracticable to sustain more than one penetration, the commander generally exploits the one enjoying the greatest success. Due to their inherent flexibility and ability to rapidly mass effects, aviation forces are well-suited to the role of an exploitation force or to enable the success of the exploitation force. Because the force conducting the penetration is vulnerable to flanking attack, it must move rapidly. Follow-on forces must be close behind to secure and widen the shoulders of the breach.

FUTURE OFFENSIVE OPERATIONS

Expeditionary maneuver warfare and emerging technologies will have a major impact on how the Marine Corps will conduct offensive operations in the future. New information technologies will allow the commander to share his operational design and situational awareness with his subordinates much faster and clearer than in the past. All commanders will share a common operational picture, specifically tailored for their echelon of command. This situational awareness, coupled with a common operating picture will allow commanders to synchronize the actions of their forces, assess the effects of their operations, and make rapid adjustments to the plan. Subordinate commanders will have the same situational awareness as their commander allowing them to exercise their initiative to meet the commander's intent without waiting for direction from their higher headquarters. This increased ability to fuse information, determine its significance, and exploit the resulting opportunities will help maintain the initiative and generate tempo.

New doctrine, organizations, and training based on evolving tactics and equipment will allow commanders to mass the effects of long range fires and agile maneuver, rather than massing forces to deliver the decisive stroke. New intelligence collection and surveillance technologies will allow the commander to accurately and rapidly locate the enemy and will reduce the need to conduct costly and time-consuming movements to contact or meeting engagements. New target acquisition equipment and fire support command and control systems will increase responsiveness and enable emerging "sensor to shooter" technologies.

Expeditionary maneuver warfare and emerging technologies will enable the MAGTF commander to conduct simultaneous operations across the battlespace to defeat specific enemy capabilities. The effects of these operations will present the enemy with multiple, simultaneous dilemmas for the MAGTF commander to exploit.

CHAPTER 8

The MAGTF in the Defense

Contents

Purpose of Defensive Operations	8-3
Characteristics of MAGTF Defensive Operations	8-4
Preparation	8-7
Security	8-7
Disruption	8-7
Mass and Concentration	8-8
Flexibility	8-8
Maneuver	8-8
Operations in Depth	8-8
Organization of the Battlespace	8-10
Security Area	8-11
Main Battle Area	8-12
Rear Area	8-12
Organization of the Force	8-12
Security Forces	8-12
Main Battle Forces	8-13
Rear Area Forces	8-15
Types of Defensive Operations	8-16
Mobile Defense	8-17
Position Defense	8-20
Future Defensive Operations	8-22

"A swift and vigorous transition to attack—the flashing sword of vengeance—is the most brilliant point of the defensive."
—Carl von Clausewitz

"Counterattack is the soul of defense . . . . We wait for the moment when the enemy shall expose himself to a counterstroke, the success of which will so far cripple him as to render us relatively strong enough to pass to the offensive ourselves."
—Julian Corbett

The commander conducts defensive operations within the context of the single battle. The MAGTF conducts defensive operations, in combination with offensive operations, to defeat an enemy attack. During the early days of the Korean War (1950-53) the 1st Marine Brigade (Provisional) conducted defensive operations along the Pusan Perimeter, buying time for the 1st Marine Division to embark and deploy to Korea where it conducted an amphibious assault at Inchon to kick off the United Nation's long awaited offensive.

Defensive operations are conducted to—

- Counter surprise action by the enemy.
- Cause an enemy attack to fail.
- Gain time.
- Concentrate combat power elsewhere.
- Increase the enemy's vulnerability by forcing him to concentrate his forces.
- Attrite or fix the enemy as a prelude to offensive operations.
- Retain decisive terrain or deny a vital area to the enemy.
- Prepare to resume the offensive.

Forward deployed or early arriving combat forces may conduct defensive operations in theater to protect the force during the build-up of combat power. During initial entry, the MAGTF may not be capable of conducting offensive operations to defeat an enemy rapidly. A classic example was the defensive operations conducted by the 1st Marine Division on Guadalcanal in 1942. These defensive operations allowed aircraft to operate from Henderson Field. It was not until 4 months after the initial landing that the 2d Marine Division and Army units could conduct offensive operations to secure the island.

Initial MAGTF forces may be assigned a mission to defend follow-on forces, air bases, and seaports in the lodgment area to provide time for the joint force commander to build sufficient combat power to support future operations. Under this condition the MAGTF must ensure sufficient combat power is available to deter or defend successfully while the buildup continues.

As a supporting effort during offensive operations, the MAGTF may be assigned to conduct defensive operations such as economy-of-force missions. This could be accomplished using air assault or amphibious forces until a larger force could linkup.

In keeping with the single battle concept, the preferred method is to conduct operations simultaneously across the depth and space of the assigned battlespace; however, it is recognized this may dictate defensive operations in some areas. The MAGTF commander and his staff will continuously make recommendations to the Marine Corps component commander and joint force commander on the proper employment of Marine forces in any type of defensive operations.

PURPOSE OF DEFENSIVE OPERATIONS

The purpose of defensive operations is to defeat an enemy attack. The MAGTF defends in order to gain sufficient strength to attack. Although offensive action is generally the decisive form of combat, it may be necessary for the MAGTF to conduct defensive operations when there is a need to buy time, hold a piece of key terrain, facilitate other operations, preoccupy the enemy in one area so friendly forces can attack him in another, or erode enemy resources at a rapid rate while reinforcing friendly operations. Defensive operations require precise synchronization since the defender is constantly seeking to regain the initiative. An effective defense consists of the following:

- Combined use of fire and maneuver to blunt the enemy's momentum.
- Speed that facilitates transition of friendly forces to the offense.
- Reducing enemy options while simultaneously increasing friendly options, thereby seizing the initiative.
- Forcing unplanned enemy culmination, gaining the initiative for friendly forces, and creating opportunities to shift to the offensive.

While the defense can deny victory to the enemy, it rarely results in victory for the defender. In many cases, however, the defense can be stronger than the offense. For example, favorable and familiar terrain, friendly civilian populations, and interior lines may prompt a commander to assume the defense to counter the advantages held by a superior enemy force. The attacking enemy usually chooses the time and place he will strike the defender. The defender uses his advantages of prepared defensive positions, concentrated firepower, obstacles, and barriers to slow the attacker's advance and disrupt the flow of his assault. Marines exploited these advantages in the defense of the Khe Sanh Combat Base, Republic of South Vietnam, during the Tet Offensive of 1968. Using aggressive defensive tactics and well-placed obstacles that were supported by responsive and continuous fires, the 26th Marines (Reinforced) destroyed two North Vietnamese Army divisions.

While on the defense, the commander conducts shaping actions, such as attacking enemy forces echeloned in depth, essential enemy sustainment capabilities, or moves his own forces and builds up fuel and ammunition to support future offensive operations. These shaping actions help to set the conditions for decisive action in the defense such as the defeat of the enemy's main effort, destruction of a critical enemy command and control node, or a counterattack as the force transitions to offensive operations.

CHARACTERISTICS OF MAGTF DEFENSIVE OPERATIONS

The objective of the defense is to force the enemy to reach his culminating point without achieving his objectives, to rapidly gain and maintain the initiative for friendly forces, and to create opportunities to shift to the offense. The integrity of the defense depends on maneuver and counterattack, as well as on the successful defense of key positions. Early identification of the enemy's committed units and direction of attack allows the defense time to react. Security forces, intelligence units, special operations forces, and aviation elements conducting deep operations will be the MAGTF's first sources of this information.

Command and control in the defense differs from the offense. Defensive operations require closer coordination, thus commanders tend to monitor the battle in more detail. Situational awareness and assessment are difficult, making identification of conditions for the resumption of the offense equally difficult.

During the defense, commanders shift their main effort to contain the enemy's attack until they can take the initiative themselves. This requires the adjustment of sectors, shifting priority of fires, repeated commitment and reconstitution of reserves, and modification of the original plan. To deny the enemy passage through a vital area, commanders may order a force to occupy a defensive position on key terrain. They also might leave a unit in position behind the enemy or give it a mission that entails a high risk of entrapment. During operations in a noncontiguous AO, units will routinely be separated from adjacent units and may be encircled by the enemy. Defending units may be unintentionally cut off from friendly forces. Whenever an unintentional encirclement occurs, the encircled commander who understands his mission and his higher commander's intent can continue contributing to his higher commander's mission.

An encircled force acts rapidly to preserve itself. The senior commander assumes control of all encircled elements and assesses the all-around defensive posture of the force. He decides whether the next higher commander wants the force to break out or to defend its position. He reorganizes and consolidates expeditiously. If the force is free to break out, it should do so before the enemy has time to block escape routes. Breaking out might mean movement of the entire encircled force, where one part is attacking and the other defending. The entire formation moves through planned escape routes created by the attacking force. If the force cannot break out, the senior commander continues to defend while planning for and assisting in a linkup with a relieving force.

Reserves preserve the commander's flexibility and provide the offensive capability of the defense. They provide the source of combat power that commanders can commit at the decisive moment. The reserve must have the mobility and striking power required to quickly isolate and defeat breakthroughs and flanking attempts. It must be able to seize and exploit fleeting opportunities in a powerful manner to throw the enemy's overall offensive off balance. The commander must organize his reserve so it can repeatedly attack, regroup, move, and attack again. Commanders may use reserves to counterattack the enemy's main effort to expedite his defeat, or they may elect to exploit enemy vulnerabilities, such as exposed flanks or support units and unprotected forces in depth. Reserves also provide a hedge against uncertainty. Reserves may reinforce forward defensive operations, block penetrating enemy forces, conduct counterattacks, or react to a rear area threat. Reserves must have multiple counterattack routes and plans that anticipate enemy's scheme of maneuver.

Helicopterborne forces can respond rapidly as reserves. On suitable terrain, they can reinforce positions to the front or on a flank. In a threatened sector, they are positioned in depth and can respond to tactical emergencies. These forces are also suitable for swift attack against enemy airborne units landing in the rear area; once committed, however, they have limited mobility.

Timing is critical to counterattacks. Commanders anticipate the circumstances that require committing the reserves. At that moment, they seek to wrest the initiative from the attacker. Commanders commit their reserves with an accurate understanding of movement and deployment times. Committed too soon, reserves may not have the desired effect or may not be available later for a more dangerous contingency. Committed too late, they may be ineffective. Once commanders commit their reserves, they should immediately begin regenerating another reserve from uncommitted forces or from forces in less threatened sectors.

During battle, protection of rear areas is necessary to ensure the defender's freedom of maneuver and continuity of operations. Because fighting in the rear area can divert combat power from the main effort, commanders carefully weigh the need for such diversions against the possible consequences and prepare to take calculated risks in rear areas. To make such decisions wisely, commanders require accurate information to avoid late or inadequate responses and to guard against overreacting to exaggerated reports.

Threats to the rear area arise throughout the battle and require the repositioning of forces and facilities. When possible, defending commanders contain enemy forces in their rear areas, using a combination of passive and active defensive measures. While commanders can never lose focus on their primary objectives, they assess risks throughout their battlespace and commit combat power where necessary to preserve their ability to accomplish the mission.

Commanders use force protection measures to preserve the health, readiness, and combat capabilities of their force. They achieve the effects of protection through skillful combinations of offense and defense, maneuver and firepower, and active and passive measures. As they conduct operations, they receive protective benefits from deep and close operations as they disrupt the attacker's tempo and blind the enemy reconnaissance efforts. Defenders also employ passive measures such as camouflage, terrain masking, and operations security to frustrate the enemy's ability to find them. Commanders should remain aware that their forces are at risk. They should adjust their activities to maintain the ability to protect their forces from attack at vulnerable points.

Weapons of mass destruction present defenders with great risks. These weapons can create gaps, destroy or disable units, and obstruct the defender's maneuver. Commanders anticipate the effects of such weapons in their defensive plans. They provide for dispersed positions for forces in depth, coordinating the last-minute concentration of units on positions with multiple routes of approach and withdrawal. They also direct appropriate training and implement protective measures.

The general characteristics of MAGTF defensive operations are *preparation*, *security*, *disruption*, *mass* and *concentration*, *flexibility*, *maneuver*, and *operations in depth*.

Preparation

The MAGTF commander organizes his defenses on terrain of his choosing. He capitalizes on the advantage of fighting from prepared positions by organizing his forces for movement and mutual support. He also conducts rehearsals to include use of the reserve and counterattack forces.

The MAGTF commander organizes his defenses in depth. Depth allows the MAGTF to push reconnaissance and surveillance forward of defended positions to detect enemy movements and to deny enemy reconnaissance. Depth allows the defense to—

- Absorb enemy attacks without suffering a breakthrough.
- Provide mutually supporting defensive positions.
- Canalize enemy forces into preset engagement areas.

Security

Security preserves the combat power of the force, allowing future employment at a time and choosing of the MAGTF. MAGTF security is achieved through the judicious use of deception that denies the enemy knowledge of friendly strengths and weaknesses.

The MAGTF plans passive measures such as dispersion, camouflage, hardening of defensive sites and facilities, barrier and obstacle plans, creation of dummy installations, and the establishment of mutually supporting positions. The MAGTF plans active measures such as conducting antiarmor and air defense operations and coordinating plans for the emplacement and security of patrols, observation posts, and reaction forces. The MAGTF may also use physical means such as a covering force in the security area to delay and disrupt enemy attacks early before they can be fully coordinated.

Disruption

The MAGTF seeks to disrupt the attacker's tempo and synchronization by countering his initiative and preventing him from massing overwhelming combat power. Disruption also affects the enemy's will to continue the attack by—

- Defeating or deceiving enemy reconnaissance and surveillance.
- Separating the enemy's forces, isolating his units, and breaking up his formations so that they cannot fight as part of an integrated whole.
- Interrupting the enemy's fire support, logistics support, and command and control.

Mass and Concentration

The MAGTF masses the effects of overwhelming combat power at the point and time of choice. Mass and concentration, while facilitating local superiority at a decisive point, may mean accepting risk in other areas.

The MAGTF must consider the collective employment of fires, maneuver, security forces, and reserve forces to mitigate this risk and, if necessary, trade terrain for time in order to concentrate forces. The MAGTF must ensure fire support assets and fire support coordination are synchronized within the overall concept of defense. This includes assignment of priority of fires, coordination of the targeting process, use of target acquisition assets, and allocation of munitions.

Flexibility

Defensive operations epitomize flexible planning and agile execution. While the attacker initially decides where and when combat will take place, agility and maneuver allow the defender to strike back effectively. Flexibility enables the MAGTF to rapidly shift the main effort, thereby constantly presenting the attacker with a coordinated, well-synchronized defense. Flexibility is enhanced by coordinating and ensuring continued sustainment to the MAGTF. Sustainment not only promotes flexibility but aids in the ability of the MAGTF to maneuver, mass fires, and concentrate forces when required. The MAGTF coordinates sustainment issues such as availability of forces, AO, infrastructure, host-nation support, sustainment bases, and basing agreements with the Marine Corps component.

Maneuver

Maneuver allows the MAGTF to take full advantage of the battlespace and to mass and concentrate when desirable. Maneuver, through movement in combination with fire, allows the MAGTF to achieve a position of advantage over the enemy to accomplish the mission. It also encompasses defensive actions such as security and rear area operations.

Operations in Depth

Simultaneous application of combat power throughout the battlespace improves the chances for success while minimizing friendly casualties. Quick, violent, and simultaneous action throughout the depth of the defender's battlespace can hurt, confuse, and even paralyze an enemy just as he is most exposed and vulnerable. Such actions weaken the enemy's will and do not allow his early successes to build confidence. Operations in depth prevent the enemy from gaining momentum in the attack. Synchronization of close, rear, and deep operations facilitates MAGTF mission success.

The ability of the MAGTF to control and influence operations throughout the depth of the battlespace prevents enemy freedom of movement. Regardless of the proximity or separation of various elements, MAGTF defense is seen as a continuous whole. The MAGTF fights deep, close, and rear operations as one battle, synchronizing simultaneous operations to a single purpose—the defeat of the enemy's attack and early transition to the offense.

Deep Operations

The MAGTF designs deep operations to achieve depth and simultaneity in the defense and to secure advantages for future operations. Deep operations disrupt the enemy's movement in depth, destroy high-payoff targets vital to the attacker, and interrupt or deny vital enemy operating systems such as command, logistics or air defense at critical times. As deep operations succeed, they upset the attacker's tempo and synchronization of effects as the defender selectively suppresses or neutralizes some of the enemy's operating systems to exploit the exposed vulnerability. Individual targets in depth are only useful as they relate to achieving the commander's desired effects, which could include destruction of a critical enemy operating system such as air defense or combat service support. As the defender denies freedom of maneuver to the attacker with deep operations, he also seeks to set the terms for the friendly force's transition to offense.

Deep operations provide protection for the force as they disrupt, delay or destroy the enemy's ability to bring combat power to bear on friendly close combat forces. As with deep operations in the offense, activities in depth such as counterbattery fire focus on effects to protect the close combat operations directly. To synchronize the activities that encompass both deep and close objectives, commanders integrate and prioritize reconnaissance, intelligence, and target acquisition efforts to focus fires and maneuver at the right place and time on the battlefield.

Close Operations

Close operations are the activities of the main and supporting efforts in the defensive area to slow, canalize, and defeat the enemy's major units. The MAGTF may do this in several ways. Often, the MAGTF will fight a series of engagements to halt or defeat enemy forces. This requires designation of a main effort, synchronization to support it, and finally a shift to concentrate forces and mass effects against another threat. This may be done repeatedly. Maneuver units defend, delay, attack, and screen as part of the defensive battle. Security operations warn of the enemy's approach and attempt to harass and slow him. A covering force meets the enemy's leading forces, strips away enemy reconnaissance and security elements, reports the attacker's strength and locations, and gives the MAGTF time and space in which to react to the enemy.

Reserves conduct operations throughout the defense and may require continual regeneration. They give the MAGTF the means to seize the initiative and to preserve their flexibility; they seek to strike a decisive blow against the attacker but prepare to conduct other missions as well. They provide a hedge against uncertainty. Reserves operate best when employed to reinforce and expedite victory rather than prevent defeat.

Rear Operations

Rear operations protect the force and sustain combat operations. Successful rear operations allow the MAGTF freedom of action by preventing disruption of command and control, fire support, logistical support, and movement of reserves. Destroying or neutralizing enemy deep battle forces achieves this goal.

Enemy forces may threaten the rear during establishment of the initial lodgment and throughout operations in theater. Initially, close and rear operations overlap due to the necessity to protect the buildup of combat power. Later, deep, close, and rear operations may not be contiguous. When this situation occurs, rear operations must retain the initiative and deny freedom of action to the enemy, even if combat forces are not available. A combination of passive and active defensive measures can best accomplish this. The MAGTF assesses threat capabilities, decides where risk will be accepted, and then assigns the units necessary to protect and sustain the force. Unity of command facilitates this process.

Regardless of the proximity or separation of elements, defense of the rear is integrated with the deep and close fight. Simultaneous operations defeat the attacking enemy throughout the battlefield and allow an early transition to the offense.

To minimize the vulnerability of rear operations, command and control and support facilities in the rear area must be redundant and dispersed. Air defense elements provide defense in depth by taking positions to cover air avenues of approach and vital assets. When rear battle response forces are insufficient, tactical combat forces prepare to respond rapidly against rear area threats and prepare to move to their objectives by multiple routes.

ORGANIZATION OF THE BATTLESPACE

During defensive operations, the commander organizes his battlespace into three areas—security, main battle, and rear—in which the defending force performs specific functions. See figure 8-1. These areas can be further divided into sectors. A defensive sector is a section assigned to a subordinate commander in which he

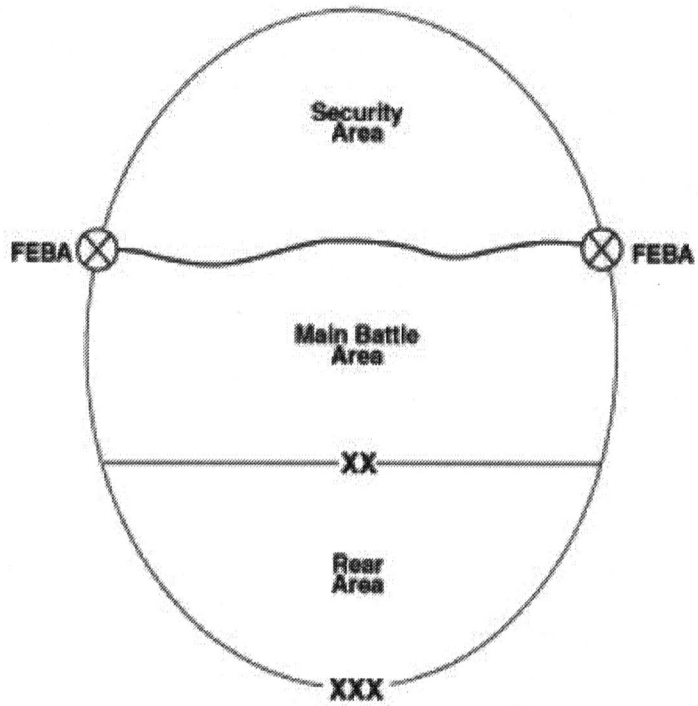

Figure 8-1. Organization of the Battlespace.

is provided the maximum latitude to accomplish assigned tasks in order to conduct defensive operations. The size and nature of a sector depends on the situation and the factors of METT-T. Commanders of defensive sectors can assign subordinates their own sector.

Security Area

The security area begins at the forward edge of the battle area (FEBA) and extends as far to the front and flanks as security forces are deployed, normally to the forward boundary of the AO. Forces in the security area conduct reconnaissance to furnish information on the enemy and delay, deceive, and disrupt the enemy. The commander adds depth to the defense by extending the security area as far forward as is tactically feasible. For more information on security operations see chapter 11.

Actions in the security area are designed to disrupt the enemy's plan of attack and cause him to prematurely deploy into attack formations. Slowing the enemy's attack enables our forces, particularly Marine aviation, to strike the enemy's critical vulnerabilities (i.e., movement, resupply, fire support, and command and control).

Main Battle Area

The main battle area is that portion of the battlespace in which the commander conducts close operations to defeat the enemy. Normally, the main battle area extends rearward from the FEBA to the rear boundary of the command's subordinate units. The commander positions forces throughout the main battle area to defeat, destroy or contain enemy assaults. Reserves may be employed in the main battle area to destroy enemy forces, reduce penetrations or regain terrain. The greater the depth of the main battle area, the greater the maneuver space for fighting the main defensive battle.

Rear Area

The rear area is that area extending forward from a command's rear boundary to the rear of the area assigned to the command's subordinate units. This area is provided primarily for the performance of combat service support functions. Rear area operations include those functions of security and sustainment required to maintain continuity of operations by the whole force. Rear area operations protect the sustainment effort as well as deny use of the rear area to the enemy. The rear area may not always be contiguous with the main battle area.

ORGANIZATION OF THE FORCE

During defensive operations, the commander organizes his force into security, main battle, and rear area forces. See figure 8-2.

Security Forces

The commander uses security forces forward of the main battle area to delay, disrupt, and provide early warning of the enemy's advance and deceive him as to the true location of the main battle area. These forces are assigned cover, guard or screen missions. Operations of security forces must be an integral part of the overall defensive plan. To ensure optimal unity of effort during security operations, a single commander is normally assigned responsibility for the conduct of operations in the security area. The composition of the security force is dependent on the factors of METT-T. A task force may be formed from the various elements of the MAGTF to conduct security operations.

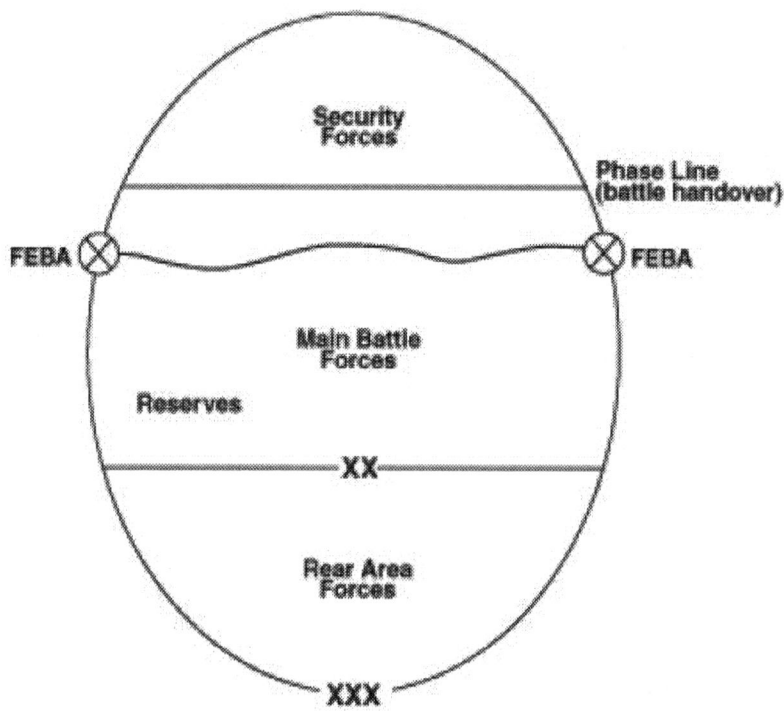

Figure 8-2. Organization of the Force.

The commander seeks to engage the enemy as far out as possible. Suppression and obscuration fires are employed to facilitate maneuver of the security force. Maximum use may be made of all fire support assets to disrupt and destroy enemy formations as they move through the security area approaching the main battle area. Obstacles and barriers are positioned to delay or canalize the enemy. They are kept under observation and covered by fires to attack him while he is halted or in the process of breaching.

Main Battle Forces

Main battle forces engage the enemy to slow, canalize, disorganize or defeat his attack. The commander positions these forces to counter the enemy's attack along the most likely or most dangerous avenue of approach. As in offensive operations, the commander weights his main effort with enough combat power and the necessary support to ensure success. When the enemy attack has been broken, the commander executes his plan to exploit any opportunity to resume the offensive.

Main battle forces engage the enemy as early as possible unless fires are withheld to prevent the loss of surprise. Commanders make maximum use of fires to destroy and disrupt enemy formations as they approach the main battle area. As the enemy closes, he is subjected to an ever-increasing volume of fires from the main battle area forces and all supporting arms. Again, obstacles and barriers are used to delay or canalize the enemy so that he is continually subjected to fires.

Combat power that can be concentrated most quickly, such as fires, is brought to bear while maneuver units move into position. The defender reacts to the enemy's main effort by reinforcing the threatened sector or allowing the enemy's main effort to penetrate into engagement areas within the main battle area to cut him off and destroy him by counterattack. Main battle forces maintain an offensive spirit throughout the battle, looking to exploit any advantageous situations.

The MAGTF commander must determine the mission, composition, and size of the reserve and counterattack forces. Reserves by definition are uncommitted forces; however, reserve forces are not uncommitted if the concept of defense depends upon their employment as a counterattack force. Counterattacking, blocking, reinforcing defending units, or reacting to rear area threats are all actions a reserve may be required to perform. The primary mission of the reserve derives directly from the concept of the defense and, therefore, the commander who established the requirement to have a reserve must approve its commitment.

A counterattack is an attack by part or all of a defending force against an attacking enemy force, for such specific purposes as regaining ground lost or cutting off and destroying enemy advance units, and with the general objective of denying to the enemy the attainment of his purpose in attacking. In many cases, the counterattack is decisive action in defensive operations. It is the commander's primary means of breaking the enemy's attack or of regaining the initiative. Once commenced, the counterattack is the main effort. Its success depends largely on surprise, speed, and boldness in execution. A separate counterattack force may be established by the commander to conduct planned counterattacks and can be made up of uncommitted or lightly engaged forces and the reserve.

The reserve is the commander's tool to influence the course of the battle at the critical time and place to exploit opportunities. The commander uses his reserve at the decisive moment in the defense and refuses to dissipate it on local emergencies. The reserve is usually located in assembly areas or forward operating bases in the main battle area. Once the reserve is committed, the commander establishes or reconstitutes a new reserve.

Reserves are organized based on the factors of METT-T. The tactical mobility of mechanized and helicopterborne forces make them well suited for use as the reserve in the defense. Mechanized reserve forces are best employed offensively. In suitable terrain, a helicopterborne reserve can react quickly to reinforce the main battle area positions or block penetrations. However, helicopterborne forces often lack the shock effect desired for counterattacks. The inherent surge capability of aviation combat forces provides the commander flexibility for reserve tasking without designating the ACE as the reserve.

Timing is critical to the employment of the reserve. As the area of probable employment of the reserve becomes apparent, the commander alerts his reserve to have it more readily available for action. When he commits his reserve, the commander must make his decision promptly and with an accurate understanding of movement factors and deployment times. If committed too soon or too late, the reserve may not have a decisive effect. The commander may choose to use security forces as part or all of his reserve after completion of their security mission. He must weigh this decision against the possibility that the security forces may suffer a loss of combat power during its security mission.

Rear Area Forces

Rear area forces protect and sustain the force's combat power. They provide for freedom of action and the continuity of logistic and command and control support. Rear area forces facilitate future operations as forces are positioned and support is marshaled to enable the transition to offensive operations. These forces should have the requisite command and control capabilities and intelligence assets to effectively employ the maneuver, fires, and combat service support forces necessary to defeat the rear area threat. Aviation forces are well suited to perform screening missions across long distances in the rear area.

The security of the rear area is provided by three levels of forces corresponding to the rear area threat level. Local security forces are employed in the rear area to repel or destroy Level I threats such as terrorists or saboteurs. These forces are normally organic to the unit, base or base cluster where they are employed. Response forces are mobile forces, with appropriate fire support designated by the area commander, employed to counter Level II threats such as enemy guerrillas or small tactical units operating in the rear area. The tactical combat force is a combat unit, with appropriate combat support and combat service support assets, that is assigned the mission of defeating Level III threats such as a large, combined arms-capable enemy force. The tactical combat force is usually located within or near the rear area where it can rapidly respond to the enemy threat.

Types of Defensive Operations

There are two fundamental types of defense: the *mobile defense* and the *position defense*. In practice, Marine commanders tend to use both types simultaneously and rarely will one type or the other be used exclusively. Mobile defense orients on the destruction of the attacking force by permitting the enemy to advance into a position that exposes him to counterattack by a mobile reserve. Position defense orients on retention of terrain by absorbing the enemy in an interlocking series of positions and destroying him largely by fires. The combination of these two types of defense can be very effective as the commander capitalizes on the advantages of each type and the strengths and capabilities of his subordinate units.

Although these descriptions convey the general pattern of each type of defense, both forms of defense employ static and dynamic elements. In mobile defenses, static defensive positions help control the depth and breadth of enemy penetration and ensure retention of ground from which to launch counterattacks. In position defenses, commanders closely integrate patrols, intelligence units, and reserve forces to cover the gaps among defensive positions, reinforcing those positions as necessary and counterattacking defensive positions as directed. Defending commanders combine both patterns, using static elements to delay, canalize, and ultimately halt the attacker, and dynamic elements (spoiling attacks and counterattacks) to strike and destroy enemy forces. The balance among these elements depends on the enemy, mission, force composition, mobility, relative combat power, and the nature of the battlefield.

The specific design and sequencing of defensive operations is an operational art largely conditioned by a thorough METT-T analysis. Doctrine allows great freedom in formulating and conducting the defense. The MAGTF commander may elect to defend well forward with strong covering forces by striking the enemy as he approaches, or he may opt to fight the decisive battle well forward within the main battle area. If the MAGTF does not have to hold a specified area or position, it may draw the enemy deep into its defenses and then strike his flanks and rear. The MAGTF commander may even choose to preempt the enemy with spoiling attacks if conditions favor such tactics.

A key characteristic of a sound defense is the ability of the commander to aggressively seek opportunities to take offensive action and wrest the initiative from the enemy. With this in mind, the decision to conduct a hasty or deliberate defense is based on the time available or the requirement to quickly resume the offense. The enemy and the mission will determine the time available.

A hasty defense is normally organized while in contact with the enemy or when contact is imminent and time available for the organization is limited. It is characterized by the improvement of natural defensive strength of the terrain by utilization of foxholes, emplacements, and obstacles. The capability to establish a robust reconnaissance effort may be limited because the defense is assumed directly from current positions. The hasty defense normally allows for only a brief leaders' reconnaissance and may entail the immediate engagement by security forces to buy time for the establishment of the defense.

Depending on the situation, it may be necessary for a commander to initially attack to seize suitable terrain on which to organize his defense. In other situations, the commander may employ a security force while withdrawing the bulk of his force some distance rearward to prepare a defense on more suitable terrain. A hasty defense is improved continuously as the situation permits, and may eventually become a deliberate defense.

A deliberate defense is normally organized when not in contact with the enemy or when contact is not imminent and time for organization is available. A deliberate defense normally includes fortifications, strong points, extensive use of barriers, and fully integrated fires. The commander normally is free to make a detailed reconnaissance of his sector, select the terrain on which to defend, and decide the best distribution of forces.

The advantage of a deliberate defense is that it allows time to plan and prepare the defense while not in contact with the enemy. A deliberate defense is characterized by a complete reconnaissance of the area to be defended by the commander and his subordinate leaders, use of key terrain, and the establishment of mutually supporting positions. The force normally has the time to create field fortifications, barriers, and emplace obstacles.

Mobile defenses sometimes rely on reserves to strike the decisive blow. They require a large, mobile, combined arms reserve. Position defenses are more likely to use reserves to block and reinforce at lower tactical levels, leaving major counterattacks to divisions and higher echelons. Regiment and battalion-level area defenses may benefit from the use of mobile reserves when such a force is available and the enemy uncovers his flanks. The actual size and composition of the reserve depend on the concept of operations.

Mobile Defense

A mobile defense is the defense of an area or position in which maneuver is used together with fire and terrain to seize the initiative from the enemy. The mobile defense destroys the attacking enemy through offensive action. The commander allocates the bulk of his combat power to mobile forces that strike the enemy where he is most vulnerable and when he least expects attack. Minimum force is placed forward to canalize, delay, disrupt, and deceive the enemy as to the actual location of our defenses. Retaining his mobile forces until the critical time and place are identified, the commander then focuses combat power in a single or series of violent and rapid counterattacks throughout the depth of the battlespace. See figure 8-3.

A mobile defense requires mobility greater than that of the attacker. Marines generate the mobility advantage necessary in the mobile defense with organic mechanized and armor forces, helicopterborne forces, and Marine aviation. The commander must have sufficient depth within the AO to allow the enemy to move into the commander's mobile defensive area. Terrain and space are traded to draw the enemy ever deeper into the defensive area, causing him to overextend his force and expose his flanks and lines of communications to attack. The success of the mobile defense often presents the opportunity to resume the offense and must be planned.

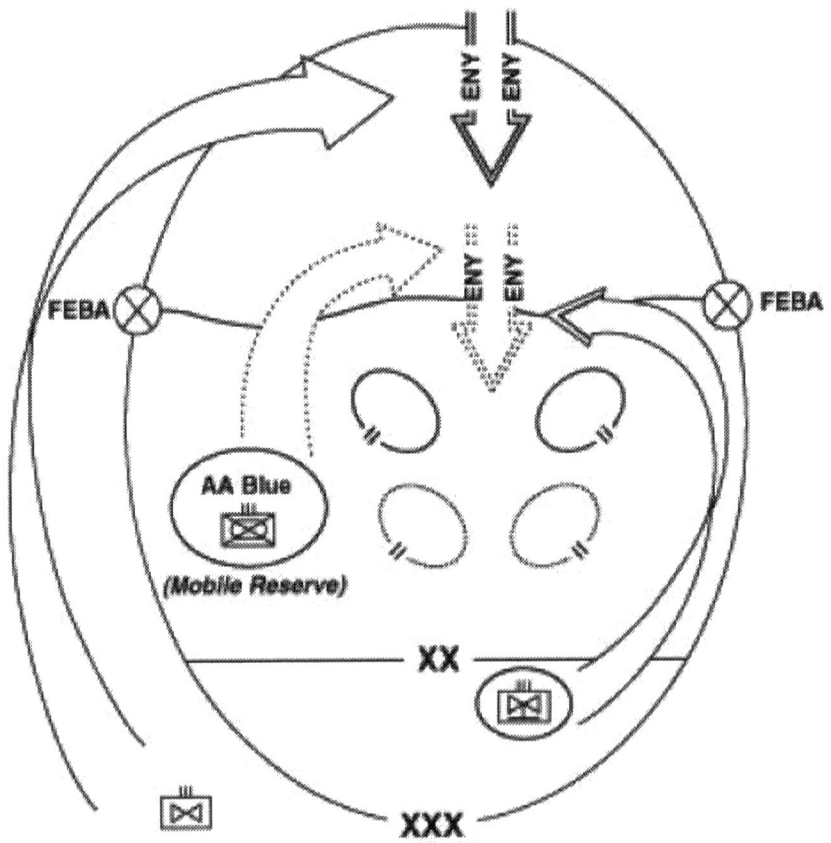

Figure 8-3. Mobile Defense.

Mobile defense orients on the destruction of the enemy force by employing a combination of fires, maneuver, offense, defense, and delay to defeat the enemy attack. Open terrain or a wide sector favors a mobile defense that orients on the enemy. The primary function of committed units in a mobile defense is to control the enemy penetration pending a counterattack by a large reserve. In mobile defense, the MAGTF commander—

- Commits minimum forces to pure defense.
- Positions maximum combat power to catch the enemy as he attempts to overcome that part of the force dedicated to the defense.
- Takes advantage of terrain in depth, obstacles, and mines, while employing firepower and maneuver to wrest the initiative from the attacker.
- Employs a strong counterattack force to strike the enemy at his most vulnerable time and place.
- Uses reconnaissance and surveillance assets to track the enemy, identifying critical enemy nodes, such as command and control, radars, logistics trains, and indirect fire support elements. These asset blind or deceive enemy critical reconnaissance and sensors, allowing less critical reconnaissance elements to draw attention to the friendly forces' secondary efforts. At the decisive moment, defenders strike simultaneously throughout the depth of the attacker's forces, assaulting him from an open flank and defeating him in detail.
- May trade terrain to divert the attention of the enemy from the main force, overextend the attacker's resources, expose his flanks, and lead him into a posture and terrain that diminishes his ability to defend against counterattack.
- Sets up large-scale counterattacks that offer opportunity to gain and retain the initiative and transition to offensive operations such as exploitation and pursuit.

Depth is required in a mobile defense to draw the enemy in and expose an exploitable weakness to counterattack. The following circumstances favor the conduct of a mobile defense:

- The defender possesses equal or greater mobility than the enemy.
- The frontage assigned exceeds the defender's capability to establish an effective position defense.
- The available battlespace allows the enemy to be drawn into an unfavorable position and exposed to attack.
- Time for preparing defensive positions is limited.
- Sufficient mechanized and aviation forces are available to allow rapid concentration of combat power.

- The enemy may employ weapons of mass destruction.
- The mission does not require denying the enemy specific terrain.

Using mobile defenses, commanders anticipate enemy penetration into the defended area and use obstacles and defended positions to shape and control such penetrations. They also use local counterattacks either to influence the enemy into entering the planned penetration area or to deceive him as to the nature of the defense. As in area defenses, static elements of a mobile defense contain the enemy in a designated area. In a mobile defense, the counterattack is strong, well-timed, and well-supported. Preferably, counterattacking forces strike against the enemy's flanks and rear rather than the front of his forces.

Position Defense

The position defense is a type of defense in which the bulk of the defending force is disposed in selected tactical positions where the decisive battle is to be fought. It denies the enemy critical terrain or facilities for a specified time. A position defense focuses on the retention of terrain by absorbing the enemy into a series of interlocked positions from which he can be destroyed, largely by fires, together with friendly maneuver. Principal reliance is placed on the ability of the forces in the defended positions to maintain their positions and to control the terrain between them. The position defense is sometimes referred to as an area defense. See figure 8-4. This defense uses battle positions, strong points, obstacles, and barriers to slow, canalize, and defeat the enemy attack. The assignment of forces within these areas and positions allows for depth and mutual support of the force.

Battle Position

A battle position is a defensive location oriented on the most likely enemy avenue of approach from which a unit may defend or attack. It can be used to deny or delay the enemy the use of certain terrain or an avenue of approach. The size of a battle position can vary with the size of the unit assigned. For ground combat units, battle positions are usually hastily occupied but should be continuously improved.

Strong Point

A strong point is a fortified defensive position designed to deny the enemy certain terrain as well as the use of an avenue of approach. It differs from a battle position in that it is designed to be occupied for an extended period of time. It is established on critical terrain and must be held for the defense to succeed. A strong point is organized for all-around defense and should have sufficient supplies and ammunition to continue to fight even if surrounded or cut off from resupply. Strong points require considerable time and engineer resources.

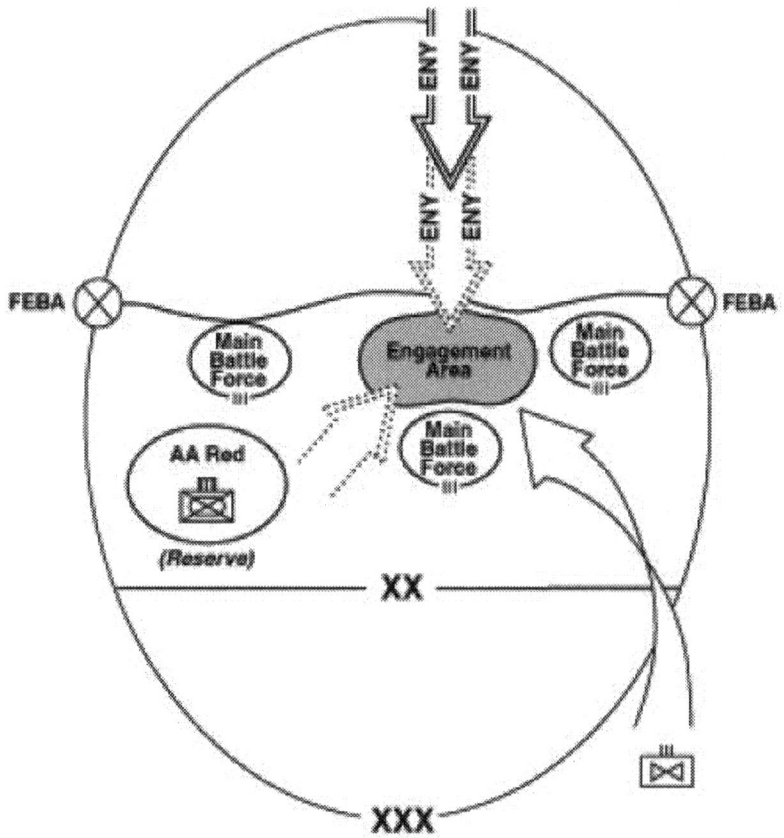

Figure 8-4. Position Defense.

Preparation of a position defense is a continuing process that ends only when the defender is ordered to give up the terrain. METT-T drives the tasks to be done and their priority, making maximum use of obstacle and barrier plans, engagement areas, and fires. Mobile defenses require considerable depth, but position defenses vary in depth according to the situation. For example, a significant obstacle to the front—such as a river, built-up area, swamp, or escarpment—favors a position defense. Such an obstacle adds to the relative combat power of the defender. Obstacles support static elements of the defense and slow or canalize the enemy in vital areas.

The commander positions the bulk of his combat power in static defensive positions and small mobile reserves. He depends on his static forces to defend their positions. His reserves are used to blunt and contain penetrations, to counterattack, and to exploit opportunities presented by the enemy. The commander also employs security forces in the position defense. The commander conducts a position defense under the following circumstances:

- The force must defend specific terrain that is militarily and politically essential.
- The defender possesses less mobility than the enemy.
- Maneuver space is limited or the terrain restricts the movement of the defending force.
- The terrain enables mutual support to the defending force.
- The depth of the battlespace is limited.
- The terrain restricts the movement of the defender.
- There is sufficient time to prepare positions.
- The employment of weapons of mass destruction by the enemy is unlikely.

In a position defense, committed forces counterattack whenever conditions are favorable. Commanders use their reserves in cooperation with static elements of their defense's battle positions and strongpoints to break the enemy's momentum and reduce his numerical advantage. As the attack develops and the enemy reveals his dispositions, reserves and fires strike at objectives in depth to break up the coordination of the attack.

FUTURE DEFENSIVE OPERATIONS

Expeditionary maneuver warfare and changes in organization, doctrine, and training will alter how the MAGTF conducts defensive operations in the future. Enhancements to information technology will provide commanders with increased flexibility in the defense.

Increasingly, the MAGTF commander will receive real-time, fused information to make better informed and more timely decisions. Highly capable precision munitions, improved unmanned aerial vehicles (UAVs), and new "sensor to shooter" technologies will increase the MAGTF commander's ability to engage an attacking force, shape the battlespace, and set conditions for decisive actions. Increased ranges of fire support systems and improved mobility of ground forces, using advanced amphibious assault vehicles and innovative aircraft, will allow the commander to mass effects on the enemy instead of massing forces that become more susceptible to enemy counteractions. The importance of the MAGTF's access to, and use of information and information systems will not go unnoticed by future adversaries. To reduce its susceptibility, the MAGTF will require defensive activities to protect and defend the information and information systems that are critical to operational success. Ultimately, these innovations will allow the MAGTF to rapidly transition from the defense to the offensive—moving directly to exploitation and pursuit.

CHAPTER 9

Other MAGTF Tactical Operations

Contents	
Retrograde	9-1
Delay	9-2
Withdrawal	9-3
Retirement	9-4
Passage of Lines	9-5
Linkup	9-6
Relief in Place	9-7
Obstacle Crossing	9-7
Breach	9-8
River Crossing	9-9
Breakout from Encirclement	9-9

"Gentlemen, we are not retreating. We are merely attacking in another direction."
—Major General O.P. Smith, USMC

"The withdrawal should be thought of as an offensive instrument, and exercises be framed to teach how the enemy can be lured into a trap, closed by a counterstroke or devastating circle of fire."
—B.H. Liddell Hart

This chapter describes some of the other tactical operations that enable the MAGTF to execute offensive and defensive operations. These tactical operations include retrograde, passage of lines, linkup, relief in place, obstacle crossing, and breakout from encirclement. These operations are planned, coordinated, synchronized, integrated, and conducted by various MAGTF elements. For example, in the conduct of a relief in place, linkup or passage of lines, the commander ordering the operation will specify responsibilities, procedures, and resolve differences in methods of execution. The higher commander must establish measures to ensure continuous and effective fire and other support during the operation.

RETROGRADE

A retrograde is any movement or maneuver to the rear or away from the enemy. It may be forced by the enemy or may be made voluntarily. Commanders combine delay, withdrawal, and retirement for offensive and defensive schemes of maneuver. All retrograde operations are conducted to improve an operational or tactical situation or prevent a worse one from developing. They—

- Reduce the enemy's offensive capabilities.
- Draw the enemy into an unfavorable situation.
- Enable combat under conditions favorable to friendly forces.
- Gain time.
- Disengage from contact with the enemy.
- Reposition forces for commitment elsewhere.
- Shorten lines of communications.

Retrograde operations will usually involve all MAGTF elements. While the GCE is normally the main effort, the ACE and the CSSE play major roles in setting the conditions for a successful retrograde. The ACE, operating from the sea or from bases beyond the reach of the enemy's artillery, interdicts enemy forces to disrupt and delay his advance, and provides close air support to ground forces in contact and assault support to move troops, equipment, and supplies away from the enemy. The CSSE continues to provide combat service support to the MAGTF and transportation to move troops, equipment, and supplies away from the enemy and to establish new combat service support facilities in the rear to support future operations.

Delay

A delay is an operation where a force under pressure trades space for time by slowing down the enemy's momentum and inflicting maximum damage on the enemy without becoming decisively engaged. Forces execute delays when they have insufficient combat power to attack or defend or when the plan calls for drawing the enemy into an area for counterattack. See figure 9-1.

The MAGTF commander may specify the amount of time to be gained or events to be accomplished by the delaying force to successfully accomplish the mission. Delays may be used in the security area, main battle area or rear area. Sufficient depth of area is required for a delay.

Commanders should plan maximum use of terrain, barriers, and obstacles. Delaying forces must remain in constant contact with the enemy to ensure that the enemy experiences continuous pressure and to prevent delaying units from being by-passed by the enemy. The delaying force should make every effort to ensure that it does not become decisively engaged.

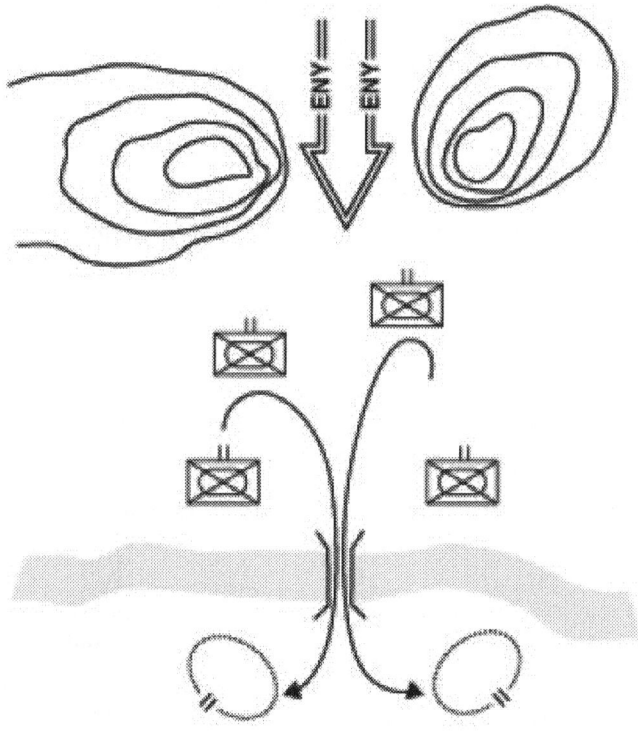

Figure 9-1. Delay.

Delays are conducted—

- When the force's strength is insufficient to defend or attack.
- To reduce the enemy's offensive capability by inflicting casualties.
- To gain time by forcing the enemy to deploy.
- To determine the strength and location of the enemy's main effort.
- When the enemy intent is not clear and the commander desires intelligence.
- To protect and provide early warning for the main battle area forces.
- To allow time to reestablish the defense.

Withdrawal

A withdrawal is a planned operation where a force in contact disengages from an enemy force. The commander's intent is to put distance between his force and the enemy. Ideally, a withdrawal is done without the enemy's knowledge or before he can prevent or disrupt it. A withdrawal is conducted—

- If the force is in danger of being defeated.
- To avoid battle under unfavorable conditions.
- To draw the enemy into terrain or a position that facilitates our offensive action.
- To allow for the reposition or redeployment of the force for employment elsewhere.

There are two types of withdrawal operations that are distinguished by the enemy's reaction to the withdrawal:

- Withdrawal under enemy pressure, in which the enemy tries to prevent the disengagement by attacking.
- Withdrawal not under enemy pressure, in which the enemy does not or cannot try to prevent the withdrawal.

Regardless of the type of withdrawal, planning considerations are the same. A prudent commander always plans to execute the withdrawal under enemy pressure. He should anticipate enemy interference by fires, direct pressure, and envelopment. If the enemy interferes, security forces will have to fight a delaying action as they move to the rear. If the enemy does not interfere, security forces disengage and withdraw on order. The more closely a unit is engaged with the enemy, the more difficult it will be to withdraw. Withdrawal will be easiest after success in combat. If a unit cannot disengage from the enemy on its own, the commander may employ other units elsewhere to reduce enemy pressure on the withdrawing unit or he may employ additional combat power at the point of withdrawal for disengagement.

In any withdrawal, the commander should attempt to deceive the enemy about his intention to withdraw. Emphasis is placed on speed and surprise. Withdrawing during periods of reduced visibility facilitates disengagement from the enemy and conceals movement to a degree but also may make control more difficult. Due to the inherent difficulties of conducting a withdrawal and the likely adverse situation, the commander must have the flexibility to switch to any other type of operation such as delay, defend or attack as the situation demands.

Retirement

A retirement is an operation where a force out of contact moves away from the enemy. A retirement may immediately follow a withdrawal. A retiring unit is normally protected by another unit between it and the enemy. However, this does not preclude the commander from having to establish adequate security during the movement. A retirement is largely an administrative movement. Speed, control, and security are the most important considerations. Commanders retire units to—

- Position forces for other missions.
- Adjust the defensive scheme.
- Prepare to assist the delays and withdrawals of other units.
- Deceive the enemy.

PASSAGE OF LINES

A passage of lines is an operation where a force moves forward or rearward through another force's combat positions with the intention of moving into or out of contact with the enemy. It is always conducted with another mission, such as to begin an attack, conduct an exploitation or a security force mission. A passage of lines, forward or rearward, is a complex and often dangerous operation, requiring thorough coordination. Inherent in the conduct of security operations, especially in the defense, is the requirement to execute a passage of lines. In an advance, security forces may be required to fix enemy forces in place and allow the main force to pass through in the attack. Faced with a superior enemy force or in the conduct of security operations in the defense, security forces must fall back and execute a rearward passage of lines, conducting battle handover with forces in the main battle area.

The conduct of a passage of lines involves the stationary force and the moving force. In the offense, the moving force is normally the attacking force and is organized to assume its assigned mission after the passage. The stationary force facilitates the passage and provides maximum support to the moving force. Normally, the plans and requirements of the moving force have priority. The time or circumstances when responsibility for the zone of action transfers from the stationary force to the moving force must be agreed upon by the two commanders or specified by higher authority. Normally, the attacking commander assumes responsibility at or before the time of attack. Responsibility may be transferred before the time of attack to allow the attacking commander to control any preparation fires. In this latter case, stationary force elements that are in contact at the time of the transfer must be placed under the operational control of the attacking commander. Liaison between the forces involved should be established as early as possible.

In the defense, a rearward passage of lines is normally executed when withdrawing a security force. The withdrawing force is the moving force and may pass through the stationary force en route to performing another mission or it may be integrated into the stationary unit. The common commander must specify any special command relationships and retains control of the passage. The actual transfer of responsibility for the sector normally is agreed upon by the executing commanders. This is carried out more effectively if the commanders are collocated. The withdrawing commander is responsible for identifying the last element of his command as it passes through the stationary unit. A detailed plan for mutual recognition must be prepared and carefully disseminated throughout both forces. The stationary commander reports to his senior commander when he has assumed responsibility for the sector. The withdrawing commander reports to the senior commander when his unit has completed the passage.

Due to the risks associated with a passage of lines, they are, if possible, conducted at night or during periods of reduced visibility. Risks include fratricide, exposure to enemy counteractions, and loss of control as responsibility for the sector is handed over from one force to another, and the potential of unintegrated movement of forces. Stationary and moving force commanders normally collocate their command posts to facilitate command and control of this demanding tactical operation.

LINKUP

A linkup is an operation where two friendly forces join together in a hostile area. The purpose of the linkup is to establish contact between two forces. A linkup may occur between a helicopterborne force and a force on the ground, between two converging forces or in the relief of an encircled force. The commander directing the linkup establishes the command relationships and responsibilities of the two units, during and after the linkup, to include responsibility for fire support coordination.

A linkup involves a stationary force and a moving force. If both units are moving, one is designated the stationary force and should occupy the linkup point at least temporarily to effect linkup. The commanders involved must coordinate their schemes of maneuver. They agree on primary and alternate linkup points where physical contact between the advance elements of the two units will occur. Linkup points must be easily recognizable to both units and are located where the routes of the moving force intersect the security elements of the stationary force. Commanders must carefully coordinate fire support for the safety of both units.

RELIEF IN PLACE

A relief in place is an operation in which, by direction of higher authority, all or part of a unit is replaced in an area by the incoming unit. Responsibilities of the replaced elements for the mission and the assigned zone of operations are transferred to the incoming unit. The incoming unit continues the operation as ordered. The relief must be executed in an expeditious and orderly manner. Every effort must be made to effect the relief without weakening the tactical integrity and security of the assigned area.

The outgoing commander is responsible for the defense of his sector until command is passed. The moment when command is to pass is determined by mutual agreement between the commanders involved, within the direction of higher headquarters. Both commanders should be collocated throughout the operation to facilitate the transfer of command and control. Following this transfer, the incoming commander will assume OPCON of all elements of the outgoing force that have not yet been relieved. The incoming commander will report to higher headquarters when he has assumed command.

The relief can take place simultaneously over the entire width of the sector or it can be staggered over time. If forces are relieved simultaneously across the sector, less time is required but greater congestion may be created. The readiness of the defense is reduced, and the enemy is more likely to detect the greater level of movement. By contrast, a relief staggered over time takes longer, but a larger portion of the force is prepared to conduct operations.

OBSTACLE CROSSING

An obstacle is a natural or manmade impediment to movement that usually requires specific techniques and equipment to overcome. A series of such obstacles is called a barrier. Crossing obstacles is most often required as part of the offense, although it may take place during the defense. Any obstacle can be crossed given sufficient time and resources. Crossings covered by the enemy, however, require extensive control and preparation to minimize losses from enemy action. A critical requirement in any obstacle crossing is the reduction or elimination of the effects of enemy fire covering the obstacle through the employment of maneuver and neutralizing, suppressing or obscuration fires. The goal is to cross the obstacle with minimum delay, loss of momentum, and disruption to concept of operations and casualties.

Upon encountering an obstacle, the commander can bypass the obstacle or execute a breach. Detailed intelligence is required to reveal the enemy's capability to oppose the crossing, the characteristics of the obstacle and crossing points, and the terrain on the far side. When possible, the attacker bypasses the enemy obstacles, saving time, labor, and risk to personnel and equipment. However, the commander must exercise caution since obstacles are often employed to canalize forces. A bypass route that at first appears desirable may lead into a killing zone.

Planning an obstacle crossing should include provisions for a hasty, and, failing or precluding that, a deliberate attempt. The commander's options may be limited because of restrictions on maneuver, the ability to deliver supporting fires, and the time required to move forces across or around the obstacle. The commander may conduct demonstrations and feints at locations away from the main crossing or breaching point to draw the enemy's defenses from that point.

The attacker advances to the obstacle quickly and on a broad front to increase the possibility of effecting a hasty crossing. The inherent capabilities of the ACE provide the MAGTF commander multiple options for moving on a broad front and rapidly crossing obstacles to establish security on the enemy side. The ACE can either provide combat assault transport for MAGTF units attacking directly across the obstacle or carry MAGTF units far beyond the obstacle to bypass the enemy or conduct a turning movement.

Once forces and equipment are committed to crossing, withdrawal or deviation from the initial plan is extremely difficult. During a crossing, a force is most vulnerable while astride the obstacle. After establishing units on the far side of the obstacle, the commander pushes his combat power across or through as quickly as possible.

Breach

When he cannot bypass an obstacle, the attacker attempts to breach. Breaching, the most common means of crossing an obstacle, is the employment of any available means to break through or secure a passage through an enemy defense, obstacle, minefield or fortification. The plan for breaching is based on the concept of operation on the far side. There are two types of breaching that generally correspond to hasty and deliberate attacks, with many of the same considerations, advantages, and disadvantages.

A hasty breach is the rapid creation of a route through a minefield, barrier or fortification by any expedient method. It is conducted as a continuation of the operation underway with a minimum loss of momentum. A hasty breach is characterized by speed and surprise, minimal concentration of forces, and decentralization of control and execution. Leading elements try to cross or breach using their own resources. Minimal engineer support, if any, is involved. Breaching equipment should be readily available to avoid the loss of momentum.

A deliberate breach is the creation of a lane through a minefield or a clear route through a barrier or fortification, which is systematically planned and carried out. It requires a concentration of the force to overcome the obstacle and enemy defenses on the far side. This requires extensive planning, detailed preparation, sustained supporting arms and engineer support. Control is centralized throughout. When forced to conduct a deliberate breach, the attacker may lose momentum and the initiative. A deliberate breach should only be implemented if the tactical situation does not permit a hasty breach.

River Crossing

Wide, unfordable rivers exercise considerable influence on military operations because they impose restrictions on movement and maneuver. They constitute obstacles to attack and form natural lines of resistance for defense. The strength of a river as an obstacle increases with width, depth, and velocity of the current. A river crossing is an operation required before ground combat power can be projected and sustained across a water obstacle. Like an amphibious operation, it is a centrally planned offensive operation that requires the thoughtful allocation of resources and control measures. The primary concern is the rapid buildup of combat power on the far side to continue offensive operations.

A hasty crossing is crossing an inland water obstacle using crossing means at hand or those readily available and made without pausing for elaborate preparations. Preferably, a hasty crossing is conducted by seizing an intact crossing site.

A deliberate crossing is crossing an inland water obstacle that requires extensive planning and detailed preparations.

BREAKOUT FROM ENCIRCLEMENT

A breakout is both an offensive and a defensive operation. An encircled force normally attempts a breakout when the—

- Breakout is ordered or is within a senior commander's intent.
- Encircled force does not have sufficient relative combat power to defend itself against the enemy.
- Encircled force does not have adequate terrain to conduct its defense.
- Encircled force cannot sustain itself for any length of time or until relieved by friendly forces.

The commander must execute the breakout as soon as possible. The sooner the breakout is executed, the less time the enemy has to strengthen his position and the more organic resources and support the encircled force has available. The encircled force may receive fire support and diversions from forces outside the encirclement. Most importantly, the encircled force must maintain the momentum of the attack. If the breakout fails, the force will be more vulnerable to defeat or destruction than it was before the breakout attempt.

The encircled force normally conducts a breakout by task-organizing with a force that conducts the rupture, a main body, and a rear guard. If the commander has enough forces, he may organize separate reserve, diversionary, and supporting elements. Any of these forces may consist of aviation or ground combat units (one or both as individual elements or as task-organized combined arms teams) and appropriate combat service support organizations, based on the factors of METT-T.

The force conducting the rupture, which may consist of two-thirds of the total encircled force, is assigned the mission to penetrate the enemy's encircling position, widen the gap, and hold the shoulders of the gap until all other encircled forces have moved through. The main body follows the force conducting the rupture to maintain the momentum of the attack and secure objectives past the rupture. The main body includes the main command post and the bulk of the combat service support. The rear guard provides protection for the force conducting the rupture and the main body as they pass beyond the rupture.

CHAPTER 10

Military Operations Other Than War

Contents

Principles	**10-3**
Objective	10-4
Unity of Effort	10-4
Security	10-5
Restraint	10-5
Perseverance	10-6
Legitimacy	10-6
Arms Control	**10-7**
Combatting Terrorism	**10-7**
Department of Defense Support to Counterdrug Operations	**10-7**
Enforcement of Sanctions/Maritime Intercept Operations	**10-8**
Enforcing Exclusion Zones	**10-8**
Ensuring Freedom of Navigation and Overflight	**10-8**
Humanitarian Assistance	**10-8**
Military Support to Civil Authorities	**10-9**
Nation Assistance/Support to Counterinsurgency	**10-10**
Noncombatant Evacuation Operations	**10-11**
Peace Operations	**10-12**
Protection of Shipping	**10-13**
Recovery Operations	**10-13**
Show of Force Operations	**10-14**
Strikes and Raids	**10-14**
Support to Insurgency	**10-14**
Warfighting Functions	**10-14**
Command and Control	10-15
Maneuver	10-16
Fires	10-16
Intelligence	10-17
Logistics	10-18
Force Protection	10-19

> *"One of the dominating factors in the establishment of the mission in small war situations has been in the past, and will continue to be in the future, the civil contacts of the entire command. The satisfactory solution of problems involving civil authorities and civil population requires that all ranks be familiar with the language, the geography, and the political, social, and economic factors involved in the country in which they are operating. Poor judgment on the part of subordinates in the handling of situations involving the local civil authorities and the local inhabitants is certain to involve the commander of the force in unnecessary military difficulties and cause publicity adverse to the public interests of the United States."*
> —*Small Wars Manual* (1940 Edition)

In contrast to large-scale sustained combat operations, MOOTW focuses on deterring war, resolving conflict, promoting peace, and supporting civil authorities in response to domestic crises. The Marine Corps has a long history of successful participation in MOOTW, from restoring order and nation building in Haiti and Nicaragua from 1900 to the 1930s, to guarding the United States mail in the 1920s. Capturing lessons learned from years of experience in such operations, the Marine Corps published a *Small Wars Manual* in 1940. This seminal reference publication continues to be relevant to Marines today as they face complex and sensitive situations in a variety of operations.

The national security strategy calls for engagement with other nations and a rapid response to political crises and natural disasters to help shape the security environment throughout the world. While this engagement or response may take the form of financial or political assistance, the use of United States military forces is always an option for the National Command Authorities. Combatant commanders often rely on responsive, forward-deployed MAGTFs, such as the MEU(SOC), to promote and protect national interests within their area of responsibility. These capable forces, task-organized to meet a variety of contingencies, are usually the first forces to reach the scene and are often the precursor to larger Marine and joint forces.

The Marine Corps approach to MOOTW builds on joint doctrine to better address the expeditionary nature of these types of military operations. It links Marine Corps capabilities with the collective, coordinated use of both traditional and nontraditional elements of national power into a cohesive foreign policy tool, and focuses on the ability to be expeditionary through forward-deployed naval forces. The Marine Corps' role is to provide the means for an immediate response while serving as the foundation for follow-on forces or resources. Forward-deployed MAGTFs, with their inherent range of capabilities, are well-positioned to conduct the wide range of missions and coordination with coalition, nongovernmental organizations, and other agencies essential to success in a MOOTW environment. Through IO, including information sharing and maintaining a wide range of contacts with our allies, Marines promote trust and confidence and increase the security of our allies and coalition partners. Regional engagement enhances force protection and provides an understanding of the role and preparedness of the MAGTF to respond to crises.

MOOTW may involve elements of both combat and noncombat operations in peacetime, conflict, and war. Those smaller-scale contingencies involving combat, such as peace enforcement in Haiti in 1995, Operation Urgent Fury in Grenada (1983), Operation El Dorado Canyon in Libya (1986) and Operation Just Cause in Panama (1989), may have many of the same characteristics as war, including offensive and defensive combat operations and employment of the full combat power of the MAGTF. Noncombat operations do not involve the use or threat of force and can help keep the tensions between nations below the threshold of armed conflict or war. In MOOTW, political and cultural considerations permeate planning and execution of operations at all levels of command. As in war, the goal of MOOTW is to achieve national objectives as quickly as possible.

MAGTFs conducting MOOTW are often in a support role to other governmental agencies and the United Nations. However, in certain types of MOOTW, the military may have the lead, as in small wars like Operation Urgent Fury and Operation Just Cause. MOOTW usually involve coordination with non-Department of Defense agencies and nongovernmental organizations. Although normally conducted outside of the United States, MOOTW may be conducted within the United States in support of civil authorities, as demonstrated when Marines assisted civil authorities in restoring order in Los Angeles following the 1992 riots.

Pages 10-7 through 10-14 identify the 16 types of military operations (arms control through support to insurgency) that apply to MOOTW and that a MAGTF could perform or support.

PRINCIPLES

Although primarily associated with traditional combat operations, the principles of war generally apply to MOOTW. For example, raids rely on the principles of surprise, offensive, mass, and economy of force to achieve a favorable outcome. MOOTW are conducted in a fluid environment with changing rules of engagement, always subject to rapidly transitioning from noncombat to combat operations and back again. This dynamic situation requires commanders consider not only the principles of war (based on the political situation and the nature of the crisis), but also the principles of MOOTW. JP 3-0 and JP 3-07, *Joint Doctrine for Military Operations Other Than War*, provide the following principles that apply specifically to MOOTW.

Objective

Direct every operation toward a clearly defined, decisive, and attainable objective. While defining mission success may be difficult in MOOTW, it is essential that the MAGTF remain focused on a clear and attainable objective. Specifying end state and conditions in advance helps define mission accomplishment and identify transitions between phases of the operation.

The political objectives that military objectives are based on may not specifically address the desired military end state. Commanders, from the joint force commander to the company commanders, should translate their political guidance into appropriate military objectives through a rigorous and continuous mission and threat analysis. They should carefully explain to political authorities the implications of political decisions on the capabilities of and risk to military forces. Care should be taken to avoid misunderstandings stemming from a lack of common terminology.

Changes to the initial military objectives may occur as political and military leaders gain a better understanding of the situation or because it changes. The MAGTF commander should be aware of shifts in the political objectives or in the situation itself. These changes may be very subtle, yet they may still require adjustment of the military objectives. If this adjustment is not made, the military objectives may no longer support the political objectives, legitimacy may be undermined, and force security may be compromised. Continuous assessment of the mission and end state is essential.

Unity of Effort

Seek unity of effort in every operation. This principle is derived from unity of command. It emphasizes the need to ensure all activities of the command are directed against a common objective. The often complicated command lines encountered during MOOTW place a premium on the MAGTF commander's ability to achieve consensus with his higher and adjacent commanders.

In MOOTW, achieving unity of effort is often complicated by a variety of international, foreign, and domestic military and nonmilitary participants; the lack of definitive command arrangements among them; and varying views of the objective. While the chain of command for United States military forces remains inviolate (flowing from the National Command Authorities through the combatant commander and the Marine Corps component commander to the MAGTF commander), command arrangements among coalition partners may be less well-defined and not include full command authority. The MAGTF commander must establish procedures for liaison and coordination to achieve unity of effort. Because MOOTW are often conducted at the small unit level, all levels of command must understand the informal and formal relationships.

Security

Acquire a military, political or informational advantage to prevent hostile factions from doing so. This principle enhances freedom of action by reducing vulnerability to hostile acts, influence or surprise. The uncertain nature of the situation inherent in many MOOTW, coupled with the potential for rapid change, require that operations security be an integral part of the operation. Operations security planners must consider the effect of media coverage and the possibility coverage may compromise essential security or disclose critical information.

The inherent right of self-defense against hostile acts or hostile intent applies in MOOTW; rules of engagement should not interfere with this right. Self-defense may be exercised against those persons, elements or groups hostile to the operation, such as terrorists or looters after a civil crisis or natural disaster. The MAGTF commander must be ready to counter any activity that could endanger friendly units or jeopardize the operation. All MAGTF personnel should stay alert even in a nonhostile operation with little or no perceived risk, ready to quickly transition to combat operations should circumstances dictate.

Security may also involve the protection of civilians or participating agencies and organizations. The perceived neutrality of these protected elements may be a factor in their security. Protection of a nongovernmental organization by United States military forces may create the perception that the nongovernmental organization is not neutral. Therefore, a nongovernmental organization may be reluctant to accept the United States military's protection. The MAGTF may have to balance these two potentially conflicting objectives (security and neutrality) in a diplomatic fashion through low visibility security measures that protect the civilian organization while permitting it to function in a neutral fashion.

Restraint

Apply appropriate military capability prudently. A single ill-advised act could result in significant political or military consequences. Judicious use of force is necessary. Restraint requires balancing the need for security, the conduct of the operation, and the political objectives. Excessive force antagonizes those parties involved, damaging the perception of legitimacy of United States forces while possibly enhancing that of the opposing party. The MAGTF may have to avoid the use of force and accept additional risk unless security considerations absolutely require it.

Commanders at all levels of the MAGTF should ensure their personnel know and understand the rules of engagement and are quickly informed of changes. Failure to understand and comply with established rules of engagement can result in fratricide, mission failure, and national embarrassment. Rules of engagement in MOOTW are generally more restrictive, detailed, and sensitive to political concerns than in war. Restraint is best achieved when rules of engagement issued at the beginning of an operation address most anticipated situations that may arise. Rules of engagement should be consistently reviewed and revised as necessary. Rules of engagement should be carefully scrutinized to ensure the lives and health of military personnel involved in MOOTW are not needlessly endangered.

Perseverance

Prepare for the measured, protracted application of military capability in support of strategic aims. MOOTW may require years to achieve the desired results. The underlying causes of the crisis may be elusive, making it difficult to achieve decisive resolution. The patient, resolute, and persistent pursuit of national goals and objectives, for as long as necessary to achieve them, is often a requirement for success. Commanders must be prepared to transition from periods of combat operations to noncombat operations and back again as conditions evolve.

Legitimacy

Sustain the legitimacy of the operation and of the host government, where applicable. In MOOTW, legitimacy is based on the perception of a specific audience of the legality, morality or rightness of a set of actions. This audience may be the United States public, foreign nations, the population in the AO, participating forces or potentially hostile factions. If an operation is not perceived as legitimate, the actions may not be supported and may be actively resisted. In MOOTW, legitimacy is frequently a decisive element. The prudent use of psychological operations and humanitarian and civil assistance programs assists in developing a sense of legitimacy for the supported government.

Legitimacy may depend on adherence to objectives agreed to by the international community, ensuring the action is appropriate to the situation, and fairness in dealing with various factions. It may be reinforced by restraint in the use of force, the type of forces employed, and the disciplined conduct of the forces involved. The perception of legitimacy by the United States public is strengthened if there are obvious national or humanitarian interests at stake and if there is assurance that American lives are not being needlessly or carelessly risked.

ARMS CONTROL

Arms control includes those activities by military forces to verify conventional, nuclear, biological or chemical arms control agreements; seizing or destroying weapons; dismantling or disposing of weapons and hazardous materials; and escorting deliveries of weapons. It encompasses any plan, arrangement or process controlling the numbers, types, and performance characteristics of any weapon system. These activities help reduce threats to regional security and assist in implementing arms agreements. A MAGTF may provide command and control, logistics or intelligence support to verification and inspection teams or conduct arms control inspections itself.

COMBATTING TERRORISM

Combatting terrorism involves actions to oppose terrorism including antiterrorism (defensive measures taken to reduce vulnerability of individuals and property to terrorist acts) and counterterrorism (offensive measures taken to prevent, deter, and respond to terrorism). Marine Corps forces, such as MEU(SOC), the Fleet Anti-Terrorism Security Teams, and the Chemical/Biological Incident Response Force, can participate in these kinds of operations. MAGTFs may provide secure base areas, communications, logistics, and transportation support, as well as specially tailored conventional forces to other forces conducting counterterrorism operations.

DEPARTMENT OF DEFENSE SUPPORT TO COUNTERDRUG OPERATIONS

The Department of Defense supports federal, state, and local law enforcement agencies in their efforts to disrupt the transfer of illegal drugs into the United States. Military forces assist in detecting and monitoring drug trafficking; support interdiction efforts; provide intelligence and logistic support; and integrate command, control, communications, computer, and intelligence assets dedicated to interdicting the movement of illegal drugs into the United States. Marine Corps forces frequently support Defense Department counterdrug activities, such as the Joint Task Force-Six operations along the United States-Mexico border, by providing observation, radar support, and cargo inspection.

ENFORCEMENT OF SANCTIONS/ MARITIME INTERCEPT OPERATIONS

These operations employ coercive measures to interdict the movement of designated items in or out of a nation or specified area, to compel a country or group to conform to the objectives of the nation or international body that establishes the sanctions. MAGTFs, as part of a larger naval force, can establish a barrier, detect blockade runners, and intercept and search such vessels to enforce the sanction. These operations allow authorized cargo and persons to pass through while preventing sanctioned material and persons from entering, as demonstrated during the naval blockade against Iraq throughout the 1990s.

ENFORCING EXCLUSION ZONES

These operations employ coercive measures to prohibit specified activities in a specific geographic area. An exclusion zone is established by a sanctioning body to persuade a nation or a group to modify their behavior to meet the desires of the sanctioning body or face continuing sanctions or the use or threat of force. Marine Corps forces, including Marine aviation, can enforce exclusion zones, as demonstrated in Operation Southern Watch (Iraq) and Operation Deny Flight (Bosnia).

ENSURING FREEDOM OF NAVIGATION AND OVERFLIGHT

These operations are conducted to demonstrate United States or international rights to navigate sea or air routes in accordance with international law. In 1986 United States forces conducted Operation Attain Document, a series of freedom of air and sea navigation operations, against Libya in the Gulf of Sidra. MAGTFs can conduct operations to seize and control critical chokepoints on sea lines of communication to help ensure unimpeded use of the seas. Marine aviation can assist the combatant commander provide combat air patrol (CAP) or strikes against hostile antiair missiles and guns to ensure use of air routes in accordance with international law.

HUMANITARIAN ASSISTANCE

Humanitarian assistance operations relieve or reduce the results of natural or manmade disasters that might present a serious threat to life or result in extensive damage to or loss of property. Humanitarian assistance provided by United States forces is generally limited in scope and duration. The assistance provided is designed to supplement or complement the host-nation civil authorities efforts.

The United States military provides assistance when the relief need is gravely urgent and when the humanitarian emergency overwhelms the ability of normal relief agencies to effectively respond.

Humanitarian assistance operations may be directed by the National Command Authorities when a serious international situation threatens the political or military stability of a region considered of interest to the United States or when the humanitarian situation itself may be sufficient and appropriate for employment of United States forces. The Department of State or the United States ambassador in the affected country is responsible for declaring a foreign disaster or situation that requires humanitarian assistance. The Department of State then requests Department of Defense assistance from the National Command Authorities.

Humanitarian assistance operations may cover a broad range of missions. A humanitarian assistance mission could also include securing an environment to allow humanitarian relief efforts. In 1991, 24th MEU(SOC) provided security, shelter, food, and water to the dissident Kurdish minority in northern Iraq. United States military forces participate in three basic types of humanitarian assistance operations: those coordinated by the United Nations, those where the United States acts in concert with other multinational forces or those where the United States responds unilaterally. The Marine Corps can respond rapidly to emergencies or disasters and achieve order in austere locations. This response could include providing security, logistics, engineering, medical support, and command and control and communications capabilities. Marine Corps forces can provide sea-based humanitarian assistance. The 5th MEB, during Operation Sea Angel in 1991, assisted Bangladesh in the aftermath of a devastating tropical cyclone by distributing food and medical supplies and repairing the country's transportation infrastructure.

MILITARY SUPPORT TO CIVIL AUTHORITIES

These operations provide temporary military support to domestic civil authorities when permitted by law and are normally undertaken when an emergency overwhelms the capabilities of the civil authorities. Support to civil authorities may be as diverse as temporary augmentation of government workers during strikes, restoration of law and order in the aftermath of riots or protecting life and property, and providing humanitarian relief after a natural disaster. Limitations on military forces providing support to civil authorities include the Posse Comitatus Act, which prohibits the use of federal military forces to enforce or otherwise execute laws unless expressly authorized by the Constitution or by an act of Congress. Marine Corps forces provided military support to civil authorities during disaster relief operations following Hurricane Andrew in Florida in 1992. The special purpose MAGTF established and maintained a temporary city for 2,500 displaced civilians, distributed supplies, and helped restore power to Dade County. Marines also supported civilian authorities by combatting forest fires in Idaho in 2000.

NATION ASSISTANCE/SUPPORT TO COUNTERINSURGENCY

Nation assistance is civil or military assistance, other than humanitarian assistance, rendered to a nation by military forces to promote sustained development and growth of civil institutions to promote long term regional stability. Nation assistance normally consists of security assistance, foreign internal defense, and humanitarian and civic assistance, and is integrated into the United States ambassador's country plan.

Security assistance is a series of programs the United States provides to defense equipment, military training, and other defense-related services to foreign nations in furtherance of United States national policies and objectives. The Marine Corps supports these programs by deploying mobile training teams to conduct military-to-military training. The mission of these teams is to train host-nation personnel to operate, maintain, and employ weapons and support systems or to develop a self-training capability in a particular skill. Teams may be tasked to train either military or civilian personnel, depending on host-nation requests.

Foreign internal defense programs encompass all political, economic, informational, and military support provided by the United States to another nation to assist its fight against subversion and insurgency. These programs may address other threats to a host nation's internal stability, such as civil disorder, illegal drug trafficking, and terrorism. Marine Corps military support may include training, material, advice or other assistance, including direct support and combat operations as authorized by the National Command Authorities.

Humanitarian and civic assistance programs are provided with military operations and exercises. They must fulfill military unit training requirements that incidentally provide humanitarian benefit to the host nation. Unlike humanitarian assistance operations, humanitarian and civic assistance programs are centered on planned activities to provide medical, dental, and veterinary services; construct rudimentary surface transportation systems, well drilling, and basic sanitation facilities; and rudimentary construction and repair of public facilities. Marine Corps units traditionally perform these humanitarian and civic assistance activities. In 1993 and 1994, II MEF established tent camps in Guantanamo Bay to assist Haitian and Cuban refugees. In 2000, III MEF deployed combat service support units to East Timor in the wake of civil unrest after the independence plebiscite.

NONCOMBATANT EVACUATION OPERATIONS

Noncombatant evacuation operations are the evacuation of noncombatants located in a foreign country who are faced with the threat of hostile or potentially hostile actions. They are normally conducted to evacuate United States citizens whose lives are in danger, but may also include the evacuation of United States military personnel, citizens of the host country, and third country nationals friendly to the United States, as determined by the Department of State.

The Department of State is responsible for protecting and evacuating American citizens abroad and for guarding their property. The Department of Defense advises and assists the Department of State in preparing and implementing plans for evacuating United States citizens. The United States ambassador or chief of the diplomatic mission is responsible for preparing emergency action plans that address the military evacuation of United States citizens and designated foreign nationals from a country. The conduct of military operations to assist implementation of emergency action plans is the responsibility of the geographic combatant commander.

Noncombatant evacuation operations are characterized by uncertainty. These operations may be directed without significant warning because of sudden changes in a country's government, reoriented political or military relationship with the United States, a sudden hostile threat to United States citizens from elements within or external to a foreign country or in response to a natural disaster. Noncombatant evacuation operations methods and timing are significantly influenced by diplomatic considerations. Under ideal circumstances there may be little or no opposition. However, commanders should anticipate opposition and plan the operation like any combat operation.

Noncombatant evacuation operations are similar to raids in that they involve swift insertion of a force, temporary occupation of objectives, and end with a planned withdrawal. They differ from raids in that force used is normally limited to that required to protect the evacuees and the evacuation forces. Forces penetrating foreign territory to conduct a noncombatant evacuation operation should be kept to the minimum consistent with mission accomplishment and the security of the force and the extraction and protection of evacuees. Forward-deployed MAGTFs are an ideal force to conduct these missions. Sea-basing and the task-organized nature of the MAGTF provide the combatant commander with multiple options as Marine forces did in Operations Assured Response/Quick

Response (Liberia, 1996), Nobel Obelisk (Sierra Leone, 1997), and Silver Wake (Albania, 1997). A MAGTF might conduct the entire operation itself or provide forces in a support role.

PEACE OPERATIONS

Peace operations are conducted in support of diplomatic efforts to establish and maintain peace. These operations include peace enforcement, peacekeeping, and operations in support of diplomatic efforts. Peace operations are conducted under the provisions of the United Nations Charter. The specific United Nations resolution under which a peace operation is conducted may dictate rules of engagement, use of combat power, and type of units deployed.

Peace enforcement is the application of military force or the threat of its use, usually based on international authorization or consent, to compel compliance with a generally accepted resolution or sanction. Unlike peacekeeping, peace enforcement does not require the consent of the states involved or of other parties to the conflict and the intervening force is not necessarily considered impartial. Such operations are conducted under the mandate of Chapter VII of the United Nations Charter. The purpose of peace enforcement is to maintain or restore peace and support diplomatic efforts to reach a long-term settlement. Peace enforcement operations missions include intervention operations, as well as operations to restore order, enforce sanctions, forcibly separate belligerents, and establish and supervise exclusion zones to establish an environment for truce or cease-fire. A MAGTF must deploy sufficient combat power to present a credible threat, protect the force, and conduct the full range of combat operations necessary to restore order and separate the warring factions, if necessary. Examples of peace enforcement are Operation Power Pack conducted in the Dominican Republic in 1965 and the initial phase of United States involvement in Haiti during Operation Uphold Democracy in 1995.

Peacekeeping operations are conducted with the consent of all major belligerents. They involve high levels of mutual consent and strict impartiality by the intervening force, and are authorized under Chapter VI of the United Nations Charter. They are designed to monitor and facilitate implementation of an existing truce and support diplomatic efforts to reach long-term political settlement. Forces conducting peacekeeping operations must be prepared to rapidly transition to peace enforcement operations as conditions change or consent of one or more of the belligerents is withdrawn. Preparation should include planning, task organization, equipment, and appropriate force protection measures. An example of peacekeeping is the deployment of Marine forces to

Lebanon in 1982-83. This mission began as a peacekeeping operation but evolved into a peace enforcement operation.

Operations in support of diplomatic efforts are those military actions that contribute to the furtherance of United States interests abroad. They include—

- **Preventive Diplomacy.** Preventive diplomacy involves diplomatic actions taken in advance of a predictable crisis to prevent or limit violence. Military activities that support preventive diplomacy include deployment of military forces, presence forces, and increased readiness levels to show United States resolve and ability to use force to preserve the peace.
- **Peacemaking.** Peacemaking is the process of diplomacy, mediation, negotiation, or other forms of peaceful settlement that arranges an end to a dispute, and resolves issues that led to conflict. Military activities that support peacemaking include security assistance and military-to-military relations.
- **Peace Building.** Peace building consists of post-conflict activities, primarily diplomatic, that strengthen and rebuild civil infrastructure and institutions in order to prevent a return to conflict. These activities include restoring civil authority, rebuilding physical infrastructure, and reestablishing civil institutions like schools and medical facilities.

PROTECTION OF SHIPPING

When necessary, Marine Corps forces can protect United States flag vessels, citizens, and their property embarked on United States or foreign vessels from unlawful violence in and over international waters. Protection can take the form of embarked Marines to provide security on board the vessel, ship and convoy escort and CAP by Marine aviation, and the recovery of hijacked vessels by Marine Corps security forces. MAGTFs can also protect United States shipping by eliminating threats such as a land-based antiship missiles, coastal guns, and hostile naval forces operating in the littorals.

RECOVERY OPERATIONS

Recovery operations are conducted to search for, locate, identify, rescue, and return personnel or human remains, sensitive equipment or items critical to national security. Hostile forces may oppose recovery operations. An example of a tactical recovery of aircraft and personnel is the rescue by a MEU of a downed United States Air Force pilot in Bosnia in 1995.

Show of Force Operations

Show of force operations are designed to demonstrate United States resolve and involve increased visibility of deployed forces in an attempt to defuse a specific situation that, if allowed to continue, may be detrimental to United States interests or national objectives. These operations often involve forward deployed forces such as MEU(SOC) and other MAGTFs offloading MPF assets and include multinational training exercises, rehearsals, forward deployment of units, or the buildup of forces within a theater. The Cobra Gold exercises in Thailand and Ahuas Tara exercises in Central America are examples of exercises as show of force operations.

Strikes and Raids

Strikes are offensive operations conducted to inflict damage on, seize or destroy an objective for political purposes. An example of a strike is Operation Praying Mantis where a MEU participated in the 1988 destruction of two oil platforms being used by Iran to coordinate attacks on merchant shipping.

A raid is usually a small scale operation, involving swift penetration of hostile territory to secure information, confuse the enemy or to destroy installations, followed by a planned withdrawal. Special maritime operations are principally specialized amphibious raids, including the tactical recovery of aircraft and personnel. An example of a raid is Operation El Dorado Canyon conducted against Libya in 1986, as a response to the terrorist bombing of United States servicemembers in Berlin.

Support to Insurgency

An insurgency is an organized movement aimed at the overthrow of a constituted government through the use of subversion and armed conflict. The United States may support an insurgency against a regime threatening United States interests. A MAGTF may provide logistic and training support to an insurgency but does not normally conduct combat operations against the targeted regime. An example of support to insurgency is the United States support to the Mujahadin resistance in Afghanistan during the Soviet occupation.

Warfighting Functions

Commanders and planners use the warfighting functions to achieve integrated planning and synchronized execution when planning and conducting MOOTW. See appendix E.

Command and Control

Command and control is overseen by the MAGTF commander and his subordinates and should remain flexible and promote unity of effort. No single command and control option works best for all MOOTW. MAGTFs should be flexible in modifying their standing operating procedures to meet the specific requirements of each situation.

If the MAGTF is part of a multinational force, planners must be prepared to provide increased liaison officers or advisors. Language barriers, varied cultural backgrounds, and different military capabilities and training may detract from effective coordination with multinational partners. Liaison and advisory teams must be adequately organized, staffed, trained, and equipped to overcome these detractors. Liaison teams must be provided with adequate and redundant communications means to ensure they maintain connectivity to the MAGTF. Deployment of a team may be critical to effective coordination and mission accomplishment. Marine foreign area officers possess unique capabilities in language and cross-cultural training. Their regional orientation makes them one of the principal choices to complement and support MOOTW involving multinational operations.

MAGTF planners should also consider assigning missions based on each multinational partner's capabilities. Political considerations will influence the degree of involvement for each nation. Some multinational partners may not be traditional allies of the United States. Others may harbor long-standing animosities towards other participating nations. These factors create unique interoperability, foreign disclosure, and counterintelligence issues. Early determination and resolution of these issues with partner nations during the planning process are critical for retaining the cooperation of multinational partners and ensuring they have the resources necessary to accomplish their assigned missions.

Interoperability of communications systems is critical to the success of the operation. In United States unilateral operations, command and control arrangements may vary based on coordination with military, civil authorities or federal, state, and local agencies. For example, command and control arrangements during support to United States civil authorities must be planned with unity of effort in mind, and provide communications links to appropriate United States agencies. In a disaster, routine communications may be disrupted. Civil authorities might have to rely on backup communications systems or, if civilian backup systems are disrupted, the military may have the only communications equipment available. The MAGTF should be prepared to establish communication linkages with these authorities.

Outside the United States, even on those rare occasions when the MAGTF operates unilaterally, some communications links will usually be required with local civil authorities or international agencies. Communications planning must consider the termination of United States involvement in the small war or contingency and the transfer of responsibility to another agency such as the United Nations or a nongovernmental organization. Some communications systems may have to be left behind to support the ongoing effort. This issue should be addressed early in the planning effort.

Maneuver

MOOTW encompasses both smaller-scale contingencies and other military actions that often require combat operations.

A key capability in MOOTW is to be able to act rapidly and to overwhelm the threat with a combination of potent combat forces and operational tempo. This requires the MAGTF commander to rapidly deploy the MAGTF while simultaneously employing the force in the AO. Combat power must be rapidly built up to allow the force to undertake decisive action as early in the operation as feasible. While combat operations normally require the overwhelming application of this combat power against the enemy, some MOOTW require considerable restraint in the application of deadly force. The commander and his staff must be attuned to the necessity to limit threat and noncombatant casualties and the collateral effects normally resulting from maneuver, particularly in urban areas.

Fires

A MAGTF must still plan for the use of fires in MOOTW. These fires can be lethal and nonlethal, provided by organic assets or by joint or multinational assets. The political nature and the need to maintain legitimacy makes careful mission analysis and precise use of lethal or nonlethal fires essential. MOOTW will also increasingly require the flexibility to use nonlethal fires when lethal fires are neither required nor desired. Nonlethal fires can confuse, deceive, delay or disorganize potential threat forces. Lethal fires are likely to be used sparingly, often only when necessary to protect the force. Precise planning and delivery of lethal fires are required to prevent unwanted collateral damage and avoid possible public affairs repercussions. Collateral damage can have an adverse impact on a fragile civilian infrastructure and in maintaining the support of the local population. The MAGTF must ensure it coordinates its fire support coordinating measures with its analysis of the rules of engagement.

Intelligence

Intelligence collection activities in MOOTW require a focus on the political, cultural, and economic factors that affect the situation. Information collection and analysis must often address unique and subtle problems not always encountered in war. The MAGTF will require information on many aspects of the operational environment's peoples and their cultures, politics, religion, economics, and any variances within affected groups of people. Intelligence collection should also focus quickly on transportation infrastructure in the AO, including capabilities and limitations of major seaports, airfields, and surface lines of communications.

Human intelligence (HUMINT) may provide the most useful source of information. HUMINT can supplement other intelligence sources with information not available through technical means. For example, while overhead imagery may graphically depict the number of people gathered in the town square, it cannot gauge motivations or enthusiasm of the crowds. Underdeveloped areas may not rely heavily on radio communication, denying United States forces intelligence derived through signal intercept. A HUMINT infrastructure may not be in place when United States forces arrive and should be established as quickly as possible. In coordination with theater commander and the United States country team, the MAGTF will conduct aggressive active and passive counterintelligence (CI) measures designed to neutralize threat intelligence collection activities and protect friendly information.

CI operations are vital in MOOTW. Protection of the force requires safeguarding essential elements of friendly information and countering potential threat intelligence collection efforts. Members of civilian organizations or the local populace may pass information (knowingly or unknowingly) to belligerent elements. MAGTFs must prepare for the possibility for compromise of operational information and take actions to counter the threat. Because of the possibly tenuous relationships between the United States and coalition partners during a particular small war or contingency, MAGTFs must also be alert to the possibility that covert intelligence operations may be conducted against them by a coalition partner. CI planning and operations must address this contingency in a sensitive way. Equally important is CI's role in establishing procedures and safeguards regarding the protection, handling, and release of classified or sensitive information to coalition partners and allies.

The MAGTF should exchange information with civilian organizations to benefit from their knowledge and familiarity with the culture, language, and sensitivities of the local populace. This information can be very valuable to the MAGTF as it conducts planning. However, civilian organizations may resent being considered a source of intelligence or believe that the MAGTF is seeking to recruit members of their organizations for collection efforts. MAGTF intelligence personnel should use the term "information gathering" rather than the term "intelligence." Peacekeeping operations should also use this term so that the MAGTF will be seen as conducting operations in an overt, neutral, and impartial manner. This may help the MAGTF foster better communications with other agencies, and thereby benefit from their valuable knowledge.

Logistics

Most crises requiring United States involvement will occur in developing nations that lack a robust infrastructure. Marine Corps forces can anticipate minimal host-nation support in urban and rural environments. Sea-based combat service support facilitates MOOTW, particularly in locations with a fragile infrastructure or where the security of a large support presence ashore would be at issue.

Logistics elements may be employed in quantities disproportionate to their normal military roles and in nonstandard tasks. Planners must be aware that overextending such forces may jeopardize their ability to support combat operations. Logistics elements may precede other military forces or may be the only forces deployed. Logistics personnel may be deployed to a foreign nation to support either United States or multinational forces. Logistics forces may also have a continuing responsibility, after the departure of combat forces, to support multinational forces or civilian organizations. They must be familiar with and adhere to any applicable status-of-forces agreement to which the United States is a party.

MAGTF logistics planners should analyze the capability of the host-nation economy to accommodate the logistic support required by the MAGTF or multinational forces and exercise care to limit adverse effects on the economy. Early mission analysis must also consider transportation requirements. Airfields and ports must be evaluated, particularly those in underdeveloped countries. Delay in completing the evaluation will directly affect the flow of assets into the region. Additional support forces may be required to build supporting infrastructure. These factors will impact movement of follow-on forces as well as delivery of humanitarian cargo. Procedures must be established to coordinate movement requirements and airfield slot times with other participants in the operation. Availability of fuel and other key support items may impinge on transportation support.

Medical support operations protect MAGTF personnel and support the mission by reducing the threat of uncontrolled disease problems. Medical planning should include hospitalization, preventive medicine, medical logistics, blood, medical regulating, and medical evacuation for MAGTF personnel and other individuals as designated by the commander. Preventive medicine, based on a thorough knowledge of the local environment, is a critical force protection measure that will reduce needless casualties for the MAGTF from disease in austere conditions. Medical examination of indigenous and coalition personnel and their environments will frequently provide unique operational insights for the MAGTF. The presence and use of drugs, threat development of weapons of mass destruction, and other critical evidence are often first identified or verified through this valuable information source. Preventive medicine units should be deployed and preventive medicine operations integrated into the overall operation as soon as possible. When planning for MOOTW, the potential requirement to treat the indigent population or allied military personnel must be considered. The respective capabilities of allied, civilian relief or other supporting medical forces should be evaluated before finalizing the medical support concept.

Force Protection

MOOTW involve elements of combat. Should the deterrent effect fail, these operations may escalate into war and place Marines in harm's way. The ingenuity that can be displayed by those planning and conducting attacks against MAGTF forces should not be underestimated. The impact of resulting friendly casualties on public perception and the political goals underlying the operation cannot be discounted. Appropriate force protection measures must be taken.

In addition to providing security for the MAGTF, Marine forces may need to conduct law-and-order missions during MOOTW. The MAGTF may have to plan on protecting such key host nation economic assets as telecommunications facilities, electric power plants, public health services and facilities, gas and oil storage and distribution networks, banks and finance centers, transportation networks, and water purification and distribution assets.

The local government may be unable to provide the necessary security and law and order for itself or its population. The MAGTF could be required to maintain general law and order, establish a civil defense effort, and protect the government infrastructure. While military police are the first considered for law-and-order missions, other MAGTF units may perform security missions such as patrolling, manning static guard positions, reinforcing police patrols, and crowd control.

Marine Corps forces may need to consider employment of active security measures such as hardening facilities and fortification. Buffer spaces around bases, mine detection and clearing, physical security measures such as obstacles and fences, personnel security, convoy escorts, and patrols may also be required. Protective measures against weapons of mass destruction may be increasingly necessary as these weapons proliferate and become more likely to be employed by terrorist groups.

CHAPTER 11

MAGTF Reconnaissance and Security Operations

Contents	
MAGTF Reconnaissance Assets	11-2
Command Element	11-3
Ground Combat Element	11-4
Aviation Combat Element	11-5
Combat Service Support Element	11-6
National and Theater Assets	11-6
Reconnaissance Planning	11-6
Types of Reconnaissance Missions	11-8
Route Reconnaissance	11-8
Area Reconnaissance	11-8
Zone Reconnaissance	11-9
Force-Oriented Reconnaissance	11-9
Reconnaissance Pull and Reconnaissance Push	11-9
Counterreconnaissance	11-11
Security Forces and Missions	11-12
Screen	11-13
Guard	11-14
Cover	11-16
Security Operations in Other Tactical Operations	11-17
Military Operations Other Than War	11-17

"The most certain method to uncover an enemy is to send scouts to visit suspected positions."
—*Small Wars Manual* (1940 Edition)

"Skepticism is the mother of security. Even though fools trust their enemies, prudent persons do not. The general is the principal sentinel of his army. He should always be careful of its preservation and see that it is never exposed to misfortune."
—Frederick the Great

The fog and friction of war will never allow the commander to have a perfect picture of the battlespace. However, reconnaissance operations can reduce uncertainties about an unfamiliar area and a hostile enemy who is actively trying to conceal information about his forces and intentions. Reconnaissance is an essential and continuous operation conducted to collect information and to gain and maintain contact with the enemy. Reconnaissance of some type should always precede a commitment of forces. Failure to conduct a thorough reconnaissance may cause the loss of initiative or failure to exploit fleeting opportunities. Lack of reconnaissance can result in the enemy's achieving surprise, inflicting unacceptable losses on friendly forces, and causing the failure of the mission. As part of the overall MAGTF intelligence effort, reconnaissance operations support the commander's decisionmaking process by collecting information to develop situational awareness and satisfy CCIRs.

Security operations are an essential component of all MAGTF operations. They can reduce risk by providing the MAGTF with maneuver space and reaction time. They protect the force from surprise and attempt to eliminate the unknowns in any tactical situation. Security operations prevent the enemy from collecting information on friendly forces, deceive him as to friendly capabilities and intentions, and prevent enemy forces from interfering with friendly operations. Reconnaissance operations support security operations by providing information on enemy forces, capabilities and intentions, and by denying the enemy information of friendly activities through counterreconnaissance. To succeed, a MAGTF's security operations should be integrated with its reconnaissance operations. Security operations are required during offensive, defensive and other tactical operations.

MAGTF Reconnaissance Assets

Reconnaissance is a necessary precursor to any military operation. It attempts to answer questions the commander has about the enemy that the MAGTF will fight and the battlespace in which the MAGTF will operate. The term reconnaissance describes any mission—aerial, ground or amphibious—undertaken to obtain, by visual or other detection methods, information about the activities and resources of the enemy or to secure data on the meteorological, hydrographic or geographic characteristics of a particular area. More simply, reconnaissance obtains information about the characteristics of a particular area and any known or potential enemy within it.

The commander uses reconnaissance to collect information and to gain and maintain contact with the enemy. Reconnaissance activities may range from passive surveillance to aggressive measures designed to stimulate a revealing enemy response, such as reconnaissance by fire. Passive surveillance includes systematically watching an enemy force or named area of interest; listening to an area and the activities in it to help develop intelligence needed to confirm or deny estimated threat course of action; or identifying threat critical vulnerabilities and limitations.

All MAGTF elements have reconnaissance capabilities. Each element brings to the MAGTF its own unique capabilities. Together they collect the information to plan and conduct MAGTF operations.

Command Element

The CE centralizes the planning and direction of the entire reconnaissance effort for the MAGTF. The CE can task any MAGTF element to conduct reconnaissance to satisfy the commander's information requirements. It also directly controls MAGTF-level reconnaissance assets as follows.

Radio Battalion

The radio battalion provides ground-based signals intelligence, electronic warfare, communications security monitoring, and special intelligence communications capability to support MAGTF operations. It plans and coordinates the employment of its subordinate elements, to include radio reconnaissance elements beyond the FEBA and mobile electronic warfare support system in light armored vehicles. It is the focal point for MAGTF ground-based signals intelligence operations.

Remote Sensors and Imagery Interpretation

The intelligence battalion provides remote sensor, imagery interpretation, and topographic intelligence support to MAGTF operations. In addition to the sensor control and management platoon, the force imagery interpretation unit, and the topographic platoon, the intelligence company establishes and mans the MAGTF's surveillance and reconnaissance center. It plans, executes, and monitors MAGTF reconnaissance operations.

Counterintelligence and Human Intelligence

The intelligence battalion provides HUMINT, CI, and interrogator-translator support to MAGTF operations. This support can include screening and interrogation/ debriefing of prisoners of war and persons of intelligence interest; conduct of CI force protection source operations; conduct of CI surveys and investigations; preparation of CI estimates and plans; translation of documents; and limited exploitation of captured material. In addition to the specialized CI and interrogator-translator platoons, the company employs task-organized HUMINT exploitation teams in direct support of MAGTF subordinate elements. HUMINT exploitation teams combine CI specialists and interrogator-translators in one element, thereby providing a unique and comprehensive range of CI/HUMINT services.

Force Reconnaissance

Force reconnaissance conducts distant and deep reconnaissance and surveillance in support of MAGTF operations. It uses specialized insertion, patrolling, reporting, and extraction techniques to carry out amphibious, distant, and deep reconnaissance and surveillance tasks in support of the MAGTF. Force reconnaissance maintains the capability to perform special operations-capable tasks.

Ground Combat Element

The GCE has substantial organic reconnaissance assets. Those units in contact with the enemy, especially patrols, are among the most reliable sources of information. Combat engineers are also good sources of information. These engineer units often conduct engineer reconnaissance of an area and can provide detailed reporting on lines of communications; i.e., roads, rivers, railroad lines, bridges, and obstacles to maneuver.

The mission of ground reconnaissance is to provide immediate tactical ground reconnaissance and surveillance to the GCE. Like force reconnaissance, ground reconnaissance is employed to observe and report on enemy activity and other information of military significance. Their capabilities are similar to those of force reconnaissance but ground reconnaissance does not insert units by parachute. The division reconnaissance battalion provides the ground reconnaissance assets for the GCE. The division also has other reconnaissance assets.

Light Armored Reconnaissance

Light armored reconnaissance units usually operate in forward areas or along flanks and can be relied upon to report early warning of contact with an enemy force. The Marines in each light armored vehicle are trained in information collection and reporting. These units are capable of a wide variety of missions due to their inherent mobility and organic firepower. The division light armored reconnaissance battalion provides the GCE with its light armored reconnaissance capability.

Counterbattery Radar

The counterbattery radar platoon is located within the artillery regiment's headquarters battery. It is equipped with mobile radars that detect and accurately locate enemy mortars, artillery, and rockets permitting rapid engagement with counterfire. Information on enemy order of battle and locations derived from counterbattery radar detections are reported via the GCE to the MAGTF CE.

Scout-Snipers

The scout-sniper platoon is an organic collection asset of each infantry battalion. Although the platoon can be employed in support of a myriad of tactical missions in defensive and offensive operations, they are primarily employed to provide timely surveillance and tactical data and coordinate supporting arms and close air support. The scout-sniper platoon provides the infantry battalion with extended area observation.

Aviation Combat Element

The capability of the ACE to observe the battlefield and report in near-real time gives the MAGTF commander a multidimensional capability that should be used at every opportunity. Aviation combat units can view the entire AO in depth, providing early indications and warning and reconnaissance information that can be essential to the success of the MAGTF. Each ACE aircraft (rotary- or fixed-wing), can conduct visual observation of terrain and enemy forces that it may fly over. Given the combined arms capability of the MAGTF, these aircraft can engage enemy targets immediately or direct other supporting arms against the enemy forces. The ACE manages the following reconnaissance systems.

Unmanned Aerial Vehicles

UAVs provide day-night, real-time imagery reconnaissance, surveillance, and target acquisition. Its unique capabilities can also be used to support real-time target engagement, assisting in the control of fires/supporting arms and maneuver. The UAV provides high quality video imagery for artillery or naval gunfire adjustment, battle damage assessment, and reconnaissance over land or sea. It is capable of both day and night operations using TV or forward-looking infrared cameras. UAV squadrons are under the ADCON of the ACE. The MAGTF commander retains OPCON because of the limited number of UAV assets and the critical reconnaissance capabilities they provide to the entire force. Mission tasking is exercised through the surveillance and reconnaissance center.

Advanced Tactical Airborne Reconnaissance System

The F/A-18D can be equipped with the advanced tactical airborne reconnaissance system and the Radar Upgrade Phase II with synthetic aperture radar (SAR). The advanced tactical airborne reconnaissance system is a real-time digital package providing day/night, all-weather imagery capability. The imagery collected provides sufficient detail and accuracy to permit delivery of appropriate air and ground weapons, assist with battle damage assessment, and provide tactical commanders with detailed information about the enemy's weapons, units, and disposition. Imagery resulting from collection can be digitally disseminated to the force imagery interpretation unit tactical exploitation group for exploitation, printing, and dissemination.

Electronic Reconnaissance and Warfare

Aerial electronic reconnaissance and electronic warfare is conducted using EA-6B aircraft. EA-6B aircraft also process and disseminate information from digital tape recordings obtained during electronic warfare missions to update and maintain enemy electronic order of battle information. The sensors are passive systems that require threat emitters to be active to collect.

Combat Service Support Element

The CSSE is limited in its reconnaissance capabilities, having no dedicated reconnaissance capabilities. However, it can conduct road and route reconnaissance with its engineer units, convoys, and military police. As the CSSE is often in more direct contact with the indigenous population, it can collect HUMINT unavailable to the other MAGTF elements. For example, medical battalion personnel can often provide information on health conditions and their potential impact on operations.

NATIONAL AND THEATER ASSETS

The MAGTF can draw on the full range of national, theater, joint, other Service, and allied reconnaissance assets. When made available, these capabilities will be fully integrated into MAGTF reconnaissance operations; e.g., Joint Surveillance Target Attack Radar System (JSTARS), Navy SEALS or Army signals intelligence aircraft. During forcible entry operations, the MAGTF integrates its amphibious reconnaissance capabilities with national, theater, and special operating forces.

The Marine Corps component will support the MAGTF by monitoring the status of MAGTF reconnaissance requests to national and theater entities. The component coordinates the provision of Marine intelligence liaison to the joint task force and other component intelligence elements to satisfy the MAGTF's requirements. Some MAGTF reconnaissance assets, such as the radio battalion and the CI/HUMINT company, will usually have direct connectivity with appropriate external agencies to coordinate tasking or support.

RECONNAISSANCE PLANNING

Reconnaissance supports the MAGTF commander's intent and his CCIRs. While contributing to the commander's broad situational awareness and development, reconnaissance assets tailor their efforts to support the specific CCIRs indicated by the commander's intent and subsequent unit intelligence and operations planning. Simultaneously, reconnaissance forces must remain alert to any developments that may cause the commander to reassess that intent.

MAGTF reconnaissance assets are best employed early to support the CCIRs and friendly course of action development and selection. When reconnaissance is initiated early in the planning cycle, planning and execution are driven by the flow of solid, timely information and intelligence. If reconnaissance is delayed, situation development will generally be more uncertain. In this case, planning and execution can either take place in an information vacuum or be driven by the search for such information.

Reconnaissance assets are best employed in general support. Because of the nature of warfare, MAGTF reconnaissance units will most likely be employed in rapidly developing and fluid situations. The main effort may shift quickly from one subordinate element to another. Such situations often require modifications or complete changes in reconnaissance elements' missions. The MAGTF commander and his staff are usually the most capable of determining the best use of MAGTF reconnaissance assets at any given time, to provide the necessary support, and to integrate the results of reconnaissance information with other intelligence sources. Although placing reconnaissance assets in direct support of some subordinate element or even attaching them to specific units is occasionally appropriate, in general, such support relationships make for inefficient use of specialized reconnaissance forces. Proper planning; the institution of flexible, responsive command and control and intelligence reporting procedures and networks; and clear intelligence reporting and dissemination priorities will ensure that the products of reconnaissance are shared to the maximum benefit of all potential users.

Reconnaissance requires adequate time for detailed planning and preparation. Most reconnaissance focuses on the enemy's activities and intentions to satisfy the commander's need to exploit the enemy's vulnerabilities or to attack his center of gravity. This frequently necessitates operating in and around the enemy's most critical and best defended areas. This normally requires that reconnaissance be conducted over long distances and well in advance of commencement of the operations it will support. These conditions usually dictate specialized methods of transportation, communications and information systems support, combat service support, equipment, and coordination.

TYPES OF RECONNAISSANCE MISSIONS

There are four basic types of reconnaissance: route, area, zone, and force-oriented. Each type provides specific details for mission planning and maintaining situational awareness.

An important factor in characterizing reconnaissance missions is the depth of penetration they require, which has important implications in terms of time, risk, coordination, and support requirements. The depth of penetration can be close, distant or deep.

Close reconnaissance is conducted in the area extending forward of the forward edge of the battle area to the fire support coordination line. It is directed toward determining the location, composition, disposition, capabilities, and activities of enemy committed forces. Close reconnaissance is primarily conducted by combat units manning the FEBA.

Distant reconnaissance is conducted in the far portion of the commander's area of influence. It is usually directed toward determining the location, composition, disposition and movement of supporting arms, and the reserve elements of the enemy committed forces. Distant reconnaissance is conducted beyond the FSCL to the limits of the commander's area of influence.

Deep reconnaissance is conducted beyond the commander's area of influence to the limits of the commander's area of interest (i.e., the geographic area from which information and intelligence are required to execute successful tactical operations and to plan for future operations). It is usually directed toward determining the location, composition, disposition, and movement of enemy reinforcements.

Route Reconnaissance

Route reconnaissance is a directed effort to obtain detailed information of a specified route and all terrain from which the enemy could influence movement along that route. Route reconnaissance is focused along a specific line of communication, such as a road, railway or waterway to provide new or updated information on route conditions and activities. Route reconnaissance normally precedes the movement of friendly forces. It provides detailed information about a specific route and the surrounding terrain that could be used to influence movement along that route.

Area Reconnaissance

Area reconnaissance is a directed effort to obtain detailed information on the terrain or enemy activity within a prescribed area, such as a town, ridge line, woods or other features critical to operations. Area reconnaissance can be made of a single point, such as a bridge or installation, and could include hostile headquarters, key terrain, objective areas or critical installations. Emphasis is placed on reaching the area without being detected. Hostile situations encountered en route are developed only enough to allow the reconnoitering units to report and bypass.

Zone Reconnaissance

Zone reconnaissance is a directed effort to obtain detailed information on all routes, obstacles (to include chemical or radiological contamination), terrain, and enemy forces within a zone defined by boundaries. A zone reconnaissance normally is assigned when the enemy situation is vague or when information on cross-country trafficability is desired. Zone reconnaissance concerns itself with the total integrated intelligence picture of a space defined by length and breadth. The size of the area depends on the potential for information on hostile forces, terrain, and weather in the zone; the requirements levied by the commander; and the reconnaissance forces available to exploit the intelligence value of the zone.

Force-Oriented Reconnaissance

Force-oriented reconnaissance is focused not on a geographic area but on a specific enemy organization, wherever it may be or go. Force-oriented reconnaissance concerns itself with intelligence information required about a specific enemy or target unit. Reconnaissance assets orient on that specific force, moving when necessary to observe that unit and reporting all required information (both requested and other pertinent observed and collected information).

RECONNAISSANCE PULL AND RECONNAISSANCE PUSH

A MAGTF can conduct reconnaissance using one of two basic methods—reconnaissance pull and reconnaissance push. In operations based on reconnaissance pull, information derived from reconnaissance forces guides friendly force activities. Reconnaissance elements identify the surfaces and gaps in overall enemy dispositions and permit the commander to shape the battlespace. Making rapid decisions based on the flow of information, friendly combat forces are drawn to and through the weak spots in the enemy defense and seek to quickly exploit the advantages gained. Reconnaissance pull requires early commitment of reconnaissance elements, allowance for the time to fully develop the reconnaissance picture, and a smooth flow of information from reconnaissance elements directly to higher and supported commanders and staffs in immediate need of reconnaissance data.

The landing at Tinian during World War II is an example of reconnaissance pull. Aerial and amphibious reconnaissance determined that the Japanese defenders had largely ignored the northern beaches, while focusing most of their defensive effort on most likely beaches in the southwest. The landing was changed to the northern beaches, and when coupled with a deception operation off the southern beach, resulted in a complete surprise. As Marines have done in the past, current concepts for MAGTF operations such as OMFTS and STOM depend on the ability of the MAGTF to use reconnaissance pull to determine enemy dispositions, and find or create exploitable gaps through which the MAGTF can pass while avoiding obstacles and strong points.

Reconnaissance pull requires a high tempo of intelligence operations to collect and report timely information. To sustain such operations, a reserve must be carefully maintained so that fresh reconnaissance elements are always available to support developing situations. Maintenance of a reconnaissance reserve requires adequate consideration of the time required for reconnaissance unit preparation, insertion, mission execution, extraction, and recovery. Reconnaissance pull is easiest to execute early in an operation. It is difficult to support over a lengthy period of high-tempo operations.

Operations based on reconnaissance push use reconnaissance elements more conservatively. They are often used as a tactical resource and generally with a shorter timeline. Reconnaissance push uses reconnaissance forces as the lead element of pre-planned tactical operations, detecting enemy dispositions during the movement of the entire friendly force. I MEF operations in Operation Desert Storm in 1991 were characterized by reconnaissance push. Reconnaissance forces, assisted by aggressive patrolling by combat forces, located Iraqi forces (to include the counterattack by the Iraqi 5th Mechanized Infantry Division from the Burqan oil field) forward of advancing friendly forces in enough time to prevent them from interfering with I MEF operations.

Reconnaissance pull is preferred when the MAGTF can maneuver freely and exploit the enemy weaknesses located by reconnaissance forces. It is the preferred method during offensive operations (it takes advantage of the MAGTF's inherent flexibility and reconnaissance capabilities). Reconnaissance push is more often used when the MAGTF is following a predetermined course of action, targeting and destruction of enemy forces is a priority, freedom of maneuver is limited, and usually when on the defense.

Reconnaissance, although naturally oriented towards the battlespace beyond the FEBA, should not ignore the threat to the rear area, attacking friendly command and control nodes, lines of communications, and logistics facilities. To prevent the enemy in the rear area from interfering with friendly combat operations, sufficient MAGTF reconnaissance capabilities should be devoted to the rear area.

Reconnaissance efforts must be coordinated between different echelons of command within the MAGTF as well as within the joint force to avoid duplication of effort by scarce reconnaissance assets and to facilitate turnover of reconnaissance responsibilities as units maneuver throughout the battlespace.

Reconnaissance assets may be given the additional mission of engaging enemy forces with combined arms to disrupt and delay their advance. With advances in technology, there is now an emerging capability to directly communicate between a specific MAGTF reconnaissance asset via automated data link and a fire support unit (sensor to shooter). Automatically transmitting target data from the sensor directly to the aircraft or firing unit brings combat power to bear against the enemy before the situation can change.

In MOOTW reconnaissance forces provide a broad range of capabilities from direct-action combat missions to support for disaster relief and humanitarian operations. MOOTW calls for pinpoint intelligence collection accuracy and timely reporting to support MAGTF delivery of services, fires or other support, and also usually for great restraint in the use of force. Reconnaissance operations may emphasize objectives such as the location and identification of lines of communications, services, and infrastructure to support threatened civilian populations. The CI/HUMINT capabilities of the MAGTF are exceptionally useful in the surveillance of indigenous peoples and identification and targeting of the hostile segments of the population.

COUNTERRECONNAISSANCE

At the same time that the MAGTF is conducting its reconnaissance operations the enemy will be conducting similar operations to determine the disposition of friendly forces. For example, when the enemy comes under indirect fire, they will increase their counterreconnaissance operations to locate and destroy the friendly reconnaissance elements controlling the fire. Counterreconnaissance prevents the enemy from collecting sufficient information about friendly activities to interfere with them. Counterreconnaissance consists of all measures taken to prevent hostile observation of a force, area or place. It focuses on denying the enemy access to essential elements of friendly information, information about the MAGTF in the security area, the flanks, and the rear that would further enemy objectives.

Counterreconnaissance consists of active and passive measures. Active measures detect, fix, and destroy enemy reconnaissance elements. Passive measures conceal friendly units and capabilities and deceive and confuse the enemy. There are two components of counterreconnaissance—the detection of enemy reconnaissance forces and the targeting, destruction or suppression of those reconnaissance forces so they cannot report friendly unit positions or activities. Counterreconnaissance consists of—

- Developing named areas of interest and targeted areas of interest for likely enemy reconnaissance forces.
- Conducting continuous surveillance of designated named areas of interest and targeted areas of interest.
- Executing targeting plan against enemy reconnaissance forces.
- Recovering forward security elements.

Reconnaissance operations support counterreconnaissance through collecting information on enemy reconnaissance forces, assets, and activities. Counterreconnaissance in turn supports security operations by protecting the MAGTF from enemy collection.

SECURITY FORCES AND MISSIONS

Security is inherent in all MAGTF operations. MAGTF security forces aggressively and continuously seek the enemy and reconnoiter key terrain. They conduct active reconnaissance to detect enemy movement or preparations for action and to learn as much as possible about the terrain. The ultimate goal is to determine the enemy's course of action and assist the main body in countering it. The security force uses a combination of ground patrols, observation posts, electronic warfare, and aviation assets.

Security in offensive operations is achieved by employing security elements to protect the MAGTF from unexpected attack, long range fires, and observation by the enemy. The MAGTF commander can employ a wide range of forces and capabilities to conduct security operations. This can include—

- Aviation forces to screen the main force from enemy interference during fast moving offensive operations.
- Ground forces to control, seize or retain terrain to prevent enemy observation.
- Sensors (UAVs, radar and seismic sensors) to detect the enemy or his long range fires.

Planning of security operations must consider the possibility of security forces making contact with the enemy and, if the situation dictates, include provisions for handing over the battle to the main force.

In the defense, the security force engages the enemy in the security area, screening, guarding, and covering as ordered. Normally, the commander designates the security force as his initial main effort. This force maintains contact with the enemy while falling back under pressure. At a predetermined location, normally a phase line designated as a handover line, control of the battle is transferred to the main battle force. A handover line is a control measure, preferably following easily defined terrain features, at which the responsibility for the conduct of combat operations is passed from one force to another. The transfer of control must be carefully coordinated. The main battle force supports the disengagement of the security force as it withdraws in preparation for its subsequent mission. The commander may shift the main effort to the appropriate element of the main battle force. As the enemy's advance force approaches the main battle area, execution of the defensive battle becomes increasingly decentralized.

At some point, the defending commander must plan for the enemy force breaking through the friendly security forces and approaching the main battle force. This requires transitioning friendly forces and control of the battle from security forces to the main battle force. Whenever the battle is transitioned, it requires coordination from the highest common commander.

There are three types of security missions. They vary in the degree of security provided, the forces and capabilities required, and the degree of engagement with the enemy that the commander desires. From the least degree of protection to the greatest, they are screen, guard, and cover. The forces that conduct these security missions are called a screen, a guard or a covering force. These forces may be further identified by the establishing headquarters and the location of the security force; e.g., MEF covering force, division advance guard or regimental flank screen. Security forces may consist of existing units, reinforced units or task-organized forces.

Screen

A screen observes, identifies, and reports information. It only fights in self-protection. A screen—

- Provides early warning of enemy approach.
- Gains and maintains enemy contact and reports enemy activity.
- Within capabilities, conducts counterreconnaissance.
- Within capabilities, impedes and harasses the enemy.

A screen only provides surveillance and early warning of enemy action, not physical protection. It can be employed as an economy of force measure in a low risk area because it provides security on a broad frontage with limited assets. A screen provides the least amount of protection of any security mission. It does not have the combat power to develop the situation. Aviation combat forces may be used to screen large, open areas during rapid and deep offensive operations. Terrain, weather, and the duration of the screen are critical considerations when assigning screening missions to aviation forces. The screen may operate within range of friendly artillery positioned with the ground forces because of its immediate availability. The ACE can also provide this fire support capability, but the cost in resources versus time is a factor; e.g., one section of F/A-18s may require an entire squadron for 24-hour coverage.

In offensive operations, a screening force primarily engages and destroys enemy reconnaissance elements within its capability but otherwise fights only in self-defense. It primarily uses indirect fires or close air support to destroy enemy reconnaissance elements and slow the movement of other enemy forces. The screen has the minimum combat power to provide the desired early warning while allowing the commander to retain the bulk of his combat power to execute the decisive action.

In defensive operations, security elements screen a stationary force by establishing a series of positions along a designated screen line. Positions are located to provide overlapping observation. Areas that cannot be observed from these positions are normally patrolled. Screening forces report any sightings of enemy activity and engage enemy forces with fires. Maintaining contact, the screen falls back along previously reconnoitered routes to subsequent positions. Screening forces should avoid becoming decisively engaged.

Guard

A guard protects the main force from attack, direct fire, and ground observation by fighting to gain time, while also observing and reporting information.
A guard—

- Provides early warning of enemy approach.
- Provides maneuver space to the front, flanks or rear of the force.
- Screens, attacks, defends or delays, within its capabilities, to protect the force.

A guard force protects the main body by fighting to gain time while also observing and reporting information and preventing enemy ground observation of and direct fire against the main body. A guard differs from a screen in that a guard force contains sufficient combat power to defeat, repel or fix the lead elements of an enemy ground force before they can engage the main body with direct fire.

The commander may assign guard missions to reinforced light armor units or task-organized maneuver elements, including ACE assets. A guard normally operates within range of friendly artillery positioned with the ground forces. It normally operates on a narrower front than a screen because it is expected to fight. In extreme cases, a guard may have to conduct sustained and prolonged fighting against the enemy to fulfill its primary mission to protect the force. The commander may order the guard to hold for a specified period of time.

An offensive guard may be established to the front, flanks or rear to protect the main body during the advance. Guard forces orient on the movement of the main body of the MAGTF. They provide security along specific routes of movement of the main body. They operate within the range of the main body's fire support weapons, deploying over a narrower front than a comparable-size screening force to permit concentration of combat power.

A security element guards the MAGTF in the defense by establishing a series of mutually supporting positions. The guard may establish a screen line forward of these positions. These positions immediately report any enemy contact and engage with fires at maximum range. The guard defends in place, attacks or delays to rearward positions. Routes and subsequent positions should have been previously reconnoitered.

The three types of guard forces are advance, flank, and rear.

The advance guard operates within supporting range of the main body and protects it from ground observation and direct fire.

A flank guard operates to the flank of a moving or stationary force to protect it from enemy ground observation, direct fire, and surprise attack. It must protect the entire depth of the main force's flank.

A rear guard protects the rear of the column from hostile forces. It attacks, defends, and delays as necessary, but it does not develop the situation to the point that it loses contact with the main force.

Cover

A cover operates apart from the main force to intercept, engage, delay, disorganize, and deceive the enemy before he can attack the main body. It prevents surprise during the advance. A cover—

- Gains and maintains contact with the enemy.
- Denies the enemy information about the size, strength, composition, and intention of the main force.
- Conducts counterreconnaissance and destroys enemy security forces.
- Develops the situation to determine enemy dispositions, strengths, and weaknesses.

A cover screens, guards, attacks, defends, and delays to accomplish its mission. It is a self-contained maneuver force that operates beyond the range of friendly artillery positioned with the main force. A cover may be task-organized, including aviation, artillery, tank, reconnaissance, and combat service support, to operate independently. The cover mission may be expressed in terms of time or friendly and enemy disposition; e.g., "Engage enemy forces until our main body deploys for attack."

A covering force protects the main body by fighting to gain time while also observing and reporting information and preventing enemy ground observation of and direct fire against the main body. A covering force operates outside supporting range of the main body to develop the situation. It deceives the enemy about the location of the main body while disrupting and destroying his forces. This provides the main body with the maximum early warning and reaction time. The distance forward of the main body depends on the intentions and instructions of the main body commander, the terrain, the location and strength of the enemy, and the rates of march of both the main body and the covering force. The width of the covering force area is the same as that of the main body.

Unlike a screening or guard force, a covering force is self-contained and can operate independently of the main body. A covering force or portions of it often becomes decisively engaged with enemy forces. The covering force must have substantial combat power to engage the enemy and accomplish its mission. A covering force develops the situation earlier than a screen or a guard. It fights longer and more often and defeats larger enemy forces.

While a covering force provides more security than a screen or guard, it also requires more resources. Before assigning a cover mission, the commander must ensure that he has sufficient combat power to resource a covering force and the decisive operation. When the commander lacks the resources to support both, he must assign his security force a less resource-intensive security mission, either a screen or a guard.

In the offense, a covering force is normally expected to penetrate the enemy's security forces and main defensive positions sufficiently for MAGTF main body units attacking the enemy's main defenses in depth; to identify the location and deployment of enemy forces in the main defensive positions; and to limit the ability of the enemy security forces to collect intelligence and disrupt the deployment and commitment of forces from the main body.

In the defense, a covering force conducts operations to either defend against or delay an attacking enemy force. A defensive covering force may be tasked to force the enemy to prematurely deploy and commence his attack; to identify the enemy effort; and to reduce the enemy's strength by destroying specific maneuver units and stripping away essential assets such as artillery. The defensive covering force must have mobility equal to or greater than that of the enemy.

SECURITY OPERATIONS IN OTHER TACTICAL OPERATIONS

Inherent in the conduct of security operations, especially in the defense, is the requirement to execute a passage of lines. In an advance, security forces may be required to fix enemy forces in place and allow the main body to pass through in the attack. Faced with a superior enemy force or in the conduct of security operations in the defense, security forces must fall back and execute a rearward passage of lines, handing the battle over to elements of the main body. During retrograde operations, the covering force may be used to facilitate and protect the withdrawal of the main body. At the appropriate time, the security must break contact with the enemy. To avoid decisive engagement, security forces must have the mobility and firepower to disengage. During obstacle crossings, the security force is required to protect the breaching force and prevent the enemy from interfering.

MILITARY OPERATIONS OTHER THAN WAR

Security operations in MOOTW are complicated by the requirement to extend the protection of the force to include civilians and other nongovernmental organizations, the desire for minimal casualties, and the effect of media coverage. The MAGTF commander must be ready to counter any activity that could endanger friendly units or jeopardize the operation. Even in a nonhostile operation with little or no perceived risk, all MAGTF personnel should be prepared to quickly transition to combat operations should circumstances dictate. In these situations, the mission of military forces commonly has aspects that are *preventive* in nature. That is, the MAGTF can accomplish its mission by preventing individuals or groups from carrying on undesirable activities such as rioting and looting or attacking, harassing, and otherwise threatening opponents. Sometimes hostile elements blend in with the local population of uninvolved citizens. Other times, sectors of the local population may rise against MAGTF forces and become active participants in acts of violence. Factional alignments, the level of violence, and the threat to mission accomplishment may change frequently and with little or no warning. Under such circumstances, the identity of opponents is uncertain, and the use of deadly force for purposes other than self-defense may be constrained by rules of engagement or by the judgment of the commander on the scene. See appendix E.

Nonlethal weapons expand the number of options available to commanders confronting situations where the use of deadly force poses problems. They provide flexibility by allowing the MAGTF to apply measured military force with reduced risk of serious noncombatant casualties, but still in such a manner as to provide force protection and effect compliance. Because nonlethal weapons can be employed at a lower threshold of danger, the MAGTF can respond to an evolving threat situation more rapidly. This will allow the MAGTF to retain the initiative and reduce its vulnerability. Thus, a robust nonlethal capability will assist in bringing into balance the conflicting requirements of mission accomplishment, force protection, and safety of noncombatants. It will therefore enhance the use and relevance of military force as an option while simultaneously enhancing the security of the MAGTF. The capabilities of reconnaissance forces can take on an unusual importance in MOOTW. Missions can range from direct-action combat missions to disaster relief and humanitarian assistance/disaster relief operations. Such operations call for pinpoint intelligence collection accuracy and timely reporting to support the MAGTF's delivery of services, fires or other support. They also usually require great restraint in the use of force. Reconnaissance operations may emphasize objectives such as the location and identification of lines of communications, services, and infrastructure to support threatened civilian populations.

APPENDIX A

Warfighting Functions

Contents	
Command and Control	A-1
Maneuver	A-2
Fires	A-3
Intelligence	A-3
Logistics	A-4
Force Protection	A-4

Warfighting functions are conceptual planning and execution tools used by planners and subject matter experts in each of the functional areas to produce comprehensive plans. They should not be viewed independently but as inseparable parts of a whole. Warfighting functions help the commander achieve unity of effort and build and sustain combat power. Their effective application, in concert with one another, will facilitate the planning and conduct of expeditionary operations.

COMMAND AND CONTROL

Command and control is the exercise of authority and direction over assigned and attached forces in the accomplishment of a mission. Command and control involves arranging personnel, equipment, and facilities to allow the commander to extend his influence over the force during the planning and conducting of military operations. Command and control is the overarching warfighting function that enables all of the other warfighting functions.

Command has two vital components—decisionmaking and leadership. Decisionmaking is choosing *if* to decide, then *when* and *what* to decide. It also includes recognizing the consequences of the act of deciding, and anticipating the outcomes that can be expected from the implementation of the decision. Leadership is taking responsibility for decisions; being loyal to subordinates; inspiring and directing Marines toward a purposeful end; and demonstrating physical and moral courage in the face of adversity. Command remains a very personal function. Professional competence, personality, and the will of strong commanders represent a significant part of any unit's combat power. The commander goes where he can best influence the action, where his moral and physical presence can be felt, and where his will to achieve a decision can best be expressed, understood, and acted upon. The focus of command and control is on the commander—his intent, guidance, and decisions and how he receives feedback on the results of his actions. Commanders command while staffs coordinate and make necessary control adjustments consistent with the commander's intent.

Control is inherent in command. Control allows the staff to monitor the status of the command, assess the gap between what was planned and what has been accomplished, and direct action to exploit new opportunities or correct deficiencies. Control serves its purpose if it allows the commander freedom to operate, delegate authority, lead from any critical point on the battlefield, and synchronize actions across his AO.

MANEUVER

Maneuver is the movement of forces for the purpose of gaining an advantage over the enemy in order to accomplish an objective. That advantage may be psychological, technological or temporal as well as spatial. Maneuver is movement relative to the enemy to put him at a disadvantage. It normally includes the movement of forces on the battlefield in combination with fires. Maneuver is the dynamic element of combat and the means of concentrating forces for decisive action to achieve the surprise, psychological shock, physical momentum, and moral dominance that enables smaller forces to defeat larger ones. Commanders maneuver their forces to create the conditions for tactical and operational success. Forces may maneuver in other dimensions as well. For instance, a force may also maneuver in time by increasing relative speed and operating at a faster tempo than the enemy.

Maneuver is rarely effective without firepower and force protection. Maneuver and firepower are complementary dynamics of combat. Although one might dominate a phase of the battle, the synchronized effects of both characterize combat operations. Mobility operations—such as breaching, route improvement, and bridging—preserve the freedom of maneuver of friendly forces. Countermobility operations—such as building obstacles in conjunction with fires—hinder enemy maneuver and deny mobility to enemy forces. Deception can also enhance the effectiveness of maneuver through psychological shock and surprise.

FIRES

Fires are the employment of firepower against air, ground, and sea targets. Fires delay, disrupt, degrade or destroy enemy capabilities, forces or facilities, as well as affect the enemy's will to fight. Fires include the collective and coordinated use of target acquisition systems, direct and indirect fire weapons, armed aircraft of all types, and other lethal and nonlethal means, such as electronic warfare and physical destruction. Fires are normally used in concert with maneuver and help to shape the battlespace, thus setting conditions for decisive action.

Synchronizing fires with maneuver is critical to the successful prosecution of combat operations. Commanders synchronize organic and supporting joint fire assets with their scheme of maneuver to get maximum effects of fires. Generating effective firepower against an enemy requires that organic and supporting fires be coordinated with other warfighting functions such as intelligence, maneuver, and logistics. Subordinate fire support systems and processes for determining priorities, identifying and locating targets, allocating fires assets, attacking targets, and assessing battle damage must be fully integrated. The employment of all available fires throughout the depth of the battlespace as an integrated and synchronized whole is done through the process of fire support planning, coordination, and execution.

INTELLIGENCE

Intelligence provides the commander with an understanding of the enemy and the battlespace, as well as identifying the enemy's centers of gravity and critical vulnerabilities. It assists the commander in understanding the situation, alerts him to new opportunities, and helps him assess the effects of actions upon the enemy. Intelligence drives operations and is focused on the enemy. Intelligence supports the formulation and subsequent modification of the commander's estimate of the situation by providing as accurate an image of the battlespace and the threat as possible. It is a dynamic process used to assess the current situation and confirm or deny the adoption of specific courses of action by the enemy. It helps refine the commander's understanding of the battlespace and reduces uncertainty and risk.

Intelligence provides indications and warning of potential hostile action, which prevents surprise and reduces risk from enemy actions. Intelligence supports force protection by identifying, locating, and countering an enemy's intelligence collection, sabotage, subversion, and terrorism capabilities. It also supports targeting by identifying target systems, critical nodes, and high-value targets and locating high-payoff targets. Intelligence support is critical to the planning, execution, and assessment of information operations. Finally, intelligence supports combat assessment by providing battle damage assessment, which is the timely and accurate estimate of the damage resulting from the application of military force.

LOGISTICS

Logistics encompasses all activities required to move and sustain military forces. At the tactical level, logistics is referred to as combat service support and involves arming, fueling, fixing equipment, moving, supplying, and manning, and provides personnel health services. A dependable, uninterrupted logistics system helps the commander seize and maintain the initiative. Conversely, attacking the enemy's support system can often threaten or weaken his center of gravity.

Commanders should anticipate requirements in order to push the right support forward. Tactical and operational success depends on fully integrating concepts of logistics and operations. Commanders should develop a logistics system that can react rapidly in crises or can sustain efforts to exploit tactical success. Logistics must also be prepared to support other operations, such as civil affairs. Logistics arrangements cannot be so meager that they do not meet the needs of commanders as they execute their operations, nor can they be so excessive that they overwhelm the ability of commanders to conduct operations effectively.

FORCE PROTECTION

Force protection consists of those measures taken to protect the force's fighting potential so that it can be applied at the appropriate time and place. It includes those measures the force takes to remain viable by protecting itself from the effects of enemy activities and natural occurrences. Force protection is essential to the preservation of combat power across the spectrum of operations, even in benign environments. However, since risk is an inherent condition of war, force protection does not imply over-cautiousness or the avoidance of calculated risk.

Force protection safeguards friendly centers of gravity and protects, conceals, reduces or eliminates friendly critical vulnerabilities. Hardening of facilities and fortifications of battle positions are active survivability measures. Deception, operational security, computer network defense, and dispersion—in conjunction with security operations—can increase survivability. Public affairs and civil affairs can also provide force protection by establishing a positive perception of United States forces and actions among the local population. Air defense operations provide the force with protection from enemy air and missile attack.

APPENDIX B

Principles of War

Contents	
Mass	B-1
Objective	B-2
Offensive	B-2
Security	B-2
Economy of Force	B-3
Maneuver	B-3
Unity of Command	B-4
Surprise	B-4
Simplicity	B-4

The Marine Corps' warfighting philosophy of maneuver warfare is rooted in the principles of war. These nine principles apply across the range of military operations and at the strategic, operational, and tactical levels. They are listed under the age-old acronym, "MOOSEMUSS."

The principles of war are useful aids to a commander as he considers how to accomplish his mission. They assist the commander in organizing his thinking about his mission, the enemy, the battlespace, and his forces. They should not be considered as prescriptive steps or actions that must be accomplished, but as tools to plan, execute, and assess operations. Successful application of the principles requires a commander's judgment, skill, and experience to adapt to constantly changing conditions and situations.

MASS

Concentrate the effects of combat power at the decisive place and time to achieve decisive results.

Commanders mass the effects of combat power to overwhelm the enemy and gain control of the situation. Mass applies to fires, combat support, and combat service support as well as numbers of forces. Proper use of the principle of mass, together with the other principles of war, may achieve decisive local superiority by a numerically inferior force. The decision to concentrate requires strict economy and the acceptance of risk elsewhere, particularly in view of the lethality of modern weapons that mandate rapid assembly and speedy dispersal of forces.

OBJECTIVE

Direct every military operation toward a clearly defined, decisive, and attainable objective.

The ultimate military objective of war is to defeat the enemy's forces or destroy his will to fight. The objective of each operation must contribute to this ultimate objective. Intermediate objectives must contribute quickly and economically to the purpose of the operation. The selection of an objective is based on consideration of the ultimate goal, forces available, the threat, and the AO. Every commander must clearly understand the overall mission of the higher command, his own mission, the tasks he must perform, and the reasons therefore. He considers every contemplated action in light of its direct contribution to the objective. He must clearly communicate the overall objective of the operation to his subordinates.

OFFENSIVE

Seize, retain, and exploit the initiative.

Offensive action is the decisive form of combat. Offensive action is necessary to seize, retain, and exploit the initiative and to maintain freedom of action. It allows the commander to exploit enemy weaknesses, impose his will upon the enemy, and determine the course of the battle. A defensive posture should only be a temporary expedient until the means are available to resume the offensive. Even in the conduct of a defense, the commander seeks every opportunity to seize the initiative by offensive action.

SECURITY

Never permit the enemy to acquire an unexpected advantage.

Security is those measures taken to prevent surprise, ensure freedom of action, and deny the enemy information about friendly forces, capabilities, and plans. Security is essential to the preservation of combat power across the range of military operations, even in benign environments. However, since risk is an inherent condition of war, security does not imply overcautiousness or the avoidance of calculated risk. In fact, security can often be enhanced by bold maneuver and offensive action, which deny the enemy the chance to interfere. Adequate security requires an accurate appreciation of enemy capabilities, sufficient security measures, effective reconnaissance, and continuous readiness for action.

ECONOMY OF FORCE

Allocate minimum essential combat power to secondary efforts.

Economy of force is the reciprocal of the principle of mass. The commander allocates the minimum essential combat power to secondary efforts. This requires the acceptance of prudent risks in selected areas to achieve superiority at the decisive time and location with the main effort. To devote means to unnecessary efforts or excessive means to necessary secondary efforts violates the principles of mass and objective. Economy of force measures are achieved through limited attacks, defense, deceptions or delaying actions.

MANEUVER

Place the enemy in a disadvantageous position through the flexible application of combat power.

Maneuver is the employment of forces on the battlefield through movement in combination with fires, or fire potential, to achieve a position of advantage in respect to the enemy to accomplish the mission. That advantage may be psychological, technological or temporal as well as spatial. Maneuver alone cannot usually produce decisive results; however, maneuver provides favorable conditions for closing with the enemy in decisive battle. Maneuver contributes significantly to sustaining the initiative, exploiting success, preserving freedom of action, and reducing vulnerability. Effective maneuver—in combination with mass, surprise, and economy of force—allows an inferior force to achieve decisive superiority at the necessary time and place. At all echelons, successful application of this principle requires not only fires and movement, but also flexibility of thought, plans, organization, and command and control.

Unity of Command

For every objective, ensure unity of effort under one responsible commander.

Unity of command is based on the designation of a single commander with the authority to direct and coordinate the efforts of all assigned forces in pursuit of a common objective. The goal of unity of command is unity of effort. In joint, multinational, and interagency operations where the commander may not control all elements in his AO, he seeks cooperation and builds consensus to achieve unity of effort.

Surprise

Strike the enemy at a time or place or in a manner for which he is unprepared.

The commander seeks every possible means to achieve surprise by striking the enemy at a time or place, or in a manner for which the enemy is unprepared. It is not essential that the enemy be taken unaware, but only that he become aware too late to react effectively. Factors contributing to surprise include speed, the use of unexpected forces, operating at night, effective and timely intelligence, deception, security, variation in tactics and techniques, and the use of unfavorable terrain. Surprise can decisively affect the outcome of a battle and may compensate for numerical inferiority.

Simplicity

Prepare clear, uncomplicated plans and clear, concise orders to ensure thorough understanding.

Plans should be as simple and direct as the situation and mission dictate. Direct, simple plans, and clear, concise orders reduce the chance for misunderstanding and confusion, and promote effective execution. In combat, even the simplest plan is usually difficult to execute. Other factors being equal, the simplest plan is preferred.

Multinational operations place a premium on simplicity. Language, doctrine, and cultural differences complicate military operations. Simple plans and orders minimize the confusion inherent in joint, multinational, and interagency operations.

APPENDIX C

Tactical Tasks

Contents

Enemy-Oriented Tactical Tasks	**C-2**
Ambush	C-2
Attack by Fire	C-2
Block	C-2
Breach	C-2
Bypass	C-3
Canalize	C-3
Contain	C-3
Defeat	C-3
Destroy	C-3
Disrupt	C-3
Exploit	C-3
Feint	C-3
Fix	C-4
Interdict	C-4
Neutralize	C-4
Penetrate	C-4
Reconnoiter	C-4
Rupture	C-4
Support by Fire	C-4
Terrain-Oriented Tactical Tasks	**C-4**
Clear	C-4
Control	C-5
Occupy	C-5
Reconnoiter	C-5
Retain	C-5
Secure	C-5
Seize	C-5
Friendly Force-Oriented Tactical Tasks	**C-5**
Breach	C-5
Cover	C-5
Disengage	C-5
Displace	C-6
Exfiltrate	C-6
Follow	C-6
Guard	C-6
Protect	C-6
Screen	C-6

The following commonly assigned MAGTF tactical tasks may be specified, implied or essential. They define actions the commander may take to accomplish his mission. In special circumstances, tasks may be modified to meet METT-T requirements. The commander must clearly state that he is departing from the standard meaning of these tasks. One way this can be done is by prefacing the modified task with the statement "What I mean by [modified task] is . . . ."

Tactical tasks are assigned based on capabilities. The GCE can execute all of the MAGTF's tactical tasks. The CSSE can execute those tactical tasks essential for it to provide sustainment to the MAGTF. The ACE can execute many of the MAGTF's tactical tasks but it cannot secure, seize, retain or occupy terrain without augmentation by the GCE. Weather and task duration may significantly affect the ACE's ability to execute assigned tactical tasks.

For additional information on tactical tasks, see JP 1-02; MCRP 5-12A; and MCRP 5-12C, *Marine Corps Supplement to the Department of Defense Dictionary of Military and Associated Terms*.

ENEMY-ORIENTED TACTICAL TASKS

Ambush

A surprise attack by fire from concealed positions on a moving or temporarily halted enemy.

Attack by Fire

Fires (direct and indirect) to destroy the enemy from a distance, normally used when the mission does not require or support occupation of the objective. This task is usually given to the supporting effort during offensive operations and as a counterattack option for the reserve during defensive operations. The assigning commander must specify the intent of fire—either to destroy, fix, neutralize or suppress.

Block

To deny the enemy access to a given area or to prevent enemy advance in a given direction or on an avenue of approach. It may be for a specified time. Units assigned this task may have to retain terrain.

Breach

To break through or secure a passage through a natural or enemy obstacle.

Bypass

To maneuver around an obstacle, position or enemy force to maintain the momentum of advance. Previously unreported obstacles and bypassed enemy forces are reported to higher headquarters.

Canalize

The use of existing or reinforcing obstacles or fires to restrict enemy operations to a narrow zone.

Contain

To stop, hold or surround enemy forces or to keep the enemy in a given area and prevent his withdrawing any part of his forces for use elsewhere.

Defeat

To disrupt or nullify the enemy commander's plan and overcome his will to fight, thus making him unwilling or unable to pursue his adopted course of action and yield to the friendly commander's will.

Destroy

Physically rendering an enemy force combat-ineffective unless it is reconstituted.

Disrupt

To integrate fires and obstacles to break apart an enemy's formation and tempo, interrupt his timetable or cause premature commitment or the piecemealing of his forces.

Exploit

Take full advantage of success in battle and follow up initial gains; offensive actions that usually follow a successful attack and are designed to disorganize the enemy in-depth.

Feint

An offensive action involving contact with the enemy to deceive him about the location or time of the actual main offensive action.

Fix

To prevent the enemy from moving any part of his forces, either from a specific location or for a specific period of time, by holding or surrounding them to prevent their withdrawal for use elsewhere.

Interdict

An action to divert, disrupt, delay or destroy the enemy's surface military potential before it can be used effectively against friendly forces.

Neutralize

To render the enemy or his resources ineffective or unusable.

Penetrate

To break through the enemy's defense and disrupt his defensive system.

Reconnoiter

To obtainable visual observation or other methods, information about the activities and resources of an enemy or potential enemy.

Rupture

To create a gap in enemy defensive positions quickly.

Support by Fire

Where a force engages the enemy by direct fire to support a maneuvering force using overwatch or by establishing a base of fire. The supporting force does not capture enemy forces or terrain.

TERRAIN-ORIENTED TACTICAL TASKS

Clear

The removal of enemy forces and elimination of organized resistance in an assigned zone, area or location by destroying, capturing or forcing the withdrawal of enemy forces that could interfere with the unit's ability to accomplish its mission.

Control

To maintain physical influence by occupation or range of weapon systems over the activities or access in a defined area.

Occupy

To move onto an objective, key terrain or other man-made or natural terrain area without opposition, and control the entire area.

Reconnoiter

To secure data about the meteorological, hydrographic or geographic characteristics of a particular area.

Retain

To occupy and hold a terrain feature to ensure it is free of enemy occupation or use.

Secure

To gain possession of a position or terrain feature, with or without force, and to prevent its destruction or loss by enemy action. The attacking force may or may not have to physically occupy the area.

Seize

To clear a designated area and gain control of it.

FRIENDLY FORCE-ORIENTED TACTICAL TASKS

Breach

To break through or secure a passage through a natural or friendly obstacle.

Cover

Offensive or defensive actions to protect the force.

Disengage

To break contact with the enemy and move to a point where the enemy cannot observe nor engage the unit by direct fire.

Displace

To leave one position and take another. Forces may be displaced laterally to concentrate combat power in threatened areas.

Exfiltrate

The removal of personnel or units from areas under enemy control.

Follow

The order of movement of combat, combat support, and combat service support forces in a given combat operation.

Guard

To protect the main force by fighting to gain time while also observing and reporting information.

Protect

To prevent observation, engagement or interference with a force or location.

Screen

To observe, identify and report information and only fight in self-protection.

APPENDIX D

Planning and Employment Considerations for Tactical Operations

Contents	
Offensive Operations	**D-1**
Aviation Combat Element	D-1
Ground Combat Element	D-5
Combat Service Support Element	D-9
Defensive Operations	**D-12**
Aviation Combat Element	D-12
Ground Combat Element	D-14
Combat Service Support Element	D-16
Other Tactical Operations	**D-18**
Aviation Combat Element	D-18
Ground Combat Element	D-20
Combat Service Support Element	D-22

OFFENSIVE OPERATIONS

Decisive victory rarely is the result of success gained in an initial attack; rather, it is the result of quickly and relentlessly exploiting that initial success. As specific opportunities for exploitation cannot be anticipated with certainty, the commander plans thoroughly and develops sequels based on potential outcomes of the battle. He prepares mentally for any contingency, identifying tentative concepts of operation and missions and objectives for each element of the MAGTF.

Aviation Combat Element

The aviation combat element (ACE) may conduct offensive operations to defeat, destroy or neutralize the enemy. The MAGTF must ensure that adequate battlespace is assigned to employ all the capabilities of available ACE assets. The MAGTF commander takes advantage of the ACE's capabilities—range, speed, mobility, and agility—to shape the battlespace and set conditions for decisive action. MAGTF aviation assets will be integrated into MAGTF offensive operations either as the main effort or in a supporting role.

The "Policy for Command and Control of USMC Tactical Air in Sustained Operations Ashore," found in JP 0-2, directs the MAGTF commander to provide sorties to the joint force commander for air defense, long-range interdiction, and long range reconnaissance. He must also provide sorties in excess of MAGTF direct support requirements. The MAGTF commander may task-organize aviation, ground, and combat service support units under a single commander to execute the form of offensive maneuver selected. When considering the employment of MAGTF aviation assets in the offense, planners must consider weather conditions and employment duration.

Three closely related activities occur within the MAGTF's single battle: *deep, close,* and *rear* operations. As a result, the ACE will be integral in each operation in depth to support the MAGTF's single battle.

The ACE conducts deep operations by providing fires through offensive air support (deep and close air support); force protection through antiair warfare, air reconnaissance, and electronic warfare; and support of maneuver, insertion, movement, and resupply of forces in the deep area through assault support. Security missions, such as screening, may be conducted in the deep area by the ACE.

In close operations, the ACE can be the decisive action for lasting effects on the battlefield. MAGTF commanders shape the course of the battle and can pick from a combination of the types of offensive operations and forms of maneuver to use at the critical time and place to close with and destroy the enemy. For example, commanders may fix a part of the enemy forces with aviation forces through offensive air support and then envelop using the ground combat element (GCE) to defeat the enemy. The ACE can augment the combat power of the reserve when committed by the MAGTF commander at the decisive time and place.

In rear operations, MAGTF commanders should allocate adequate resources to maintain freedom of action and continuity of operations. Aviation assets can support the force in the rear because of the range, speed, mobility, and agility. Assault support assets increase the mobility of the tactical combat force that operates in the rear area. To decrease reaction time, ACE assets may be employed as direct support assets to the rear area commander by the MAGTF commander.

Types of Offensive Operations
The ACE can conduct or support all types of offensive operations.

Movement to Contact. The initial task of the ACE is to locate the enemy by reconnoitering forward or by screening the flanks of the force. Rotary-wing aircraft are well-suited to gain, regain or maintain continuous contact with the enemy during movement to contact. Once the ACE locates the enemy it may use offensive air support to fix him. The MAGTF commander can then use the ACE to attack, to support an attack by the GCE or bypass the enemy force. During a movement to contact, aviation assets may perform a number of tasks to include:

- Reconnoiter and determine the trafficability of all high-speed routes, bridges, culverts, overpasses, underpasses, bypasses, and fords within the zone.
- Find and report all enemy forces within the zone and help determine their size, composition, and activity. The ACE is capable of establishing visual and electromagnetic contact with the enemy at extended ranges.
- Provide aviation assets for advance force, flank or rear security missions associated with the MAGTF's movement to conduct.
- Conduct screening missions.
- Provide fires and assault support for the force.

Attack. MAGTF aviation assets will be integrated into MAGTF attack operations either as the main effort or in a supporting role. During attack operations, MAGTF aviation assets may be employed in the close fight or deep against second echelon forces, enemy artillery, enemy helicopter forces, and enemy reaction forces, which could disrupt the momentum of the MAGTF attack. Operations beyond the depth of the close fight, especially when conducted in synchronization with other combined arms and joint service contributions, can break the cohesion of enemy defenses and lead to exploitation and pursuit. During attack operations, the ACE may perform a number of tasks to include:

- Disrupt, degrade or destroy specific enemy units.
- Envelop (along a specific axis) enemy forces.
- Block enemy forces.
- Conduct raids against enemy units.
- Fix enemy units.
- Screen or guard.
- Conduct counterattacks.
- Conduct feints or demonstrations.

While MAGTF aviation forces are capable of performing the tasks and/or missions listed above, they will seldom execute them alone. The MAGTF will employ forces with a variety of integrated, mutually supporting forces. An example might be the ACE attacking a second echelon enemy unit under the direction of a force reconnaissance team. To allow the aircraft to reach the target area, the GCE suppresses an enemy air defense site along the ingress route.

Exploitation. During exploitation operations, MAGTF aviation assets may be used to maintain pressure on the collapsing enemy forces. MAGTF aviation operations may be tasked to prevent the enemy from reconstituting a defense, prevent the withdrawal of enemy forces to other defensible terrain, and destroy the enemy command and control during exploitation operations. They may also be used to strike enemy, attempt to reform or provide reconnaissance in front of friendly advancing ground exploitation forces. MAGTF aerial reconnaissance gives the MAGTF commander the capability to exploit by using the greatest advantage that MAGTF aviation has to offer: range and speed.

During exploitation, the MAGTF commander assumes risk on the flanks and in the rear. He can employ aviation assets to minimize the risk by assigning the ACE to protect the flanks and can also assign direct support aviation assets to the rear area.

Pursuit. During a pursuit, the inherent speed and mobility of aviation forces are ideally suited to maintain enemy contact, develop the situation, and deliver aerial fires upon positions of enemy resistance. Since pursuit is a difficult phase of an operation to predict, ground forces may not be positioned to properly exploit the situation. Aviation forces may be moved quickly and may be tasked to find, fix, and attack fleeing enemy units; locate the enemy strike forces; and guide the GCE into attack positions or around enemy exposed flanks. The maneuverability and firepower of MAGTF aviation assets make it the optimum force to conduct pursuit operations.

Forms of Offensive Maneuver

The MAGTF commander chooses the form of maneuver that fully exploits all the dimensions of the battlespace, and that fully utilizes the capabilities of the MAGTF that best accomplishes the mission. The MAGTF commander organizes and employs the ACE to best support the chosen form of maneuver.

Envelopment. In an envelopment, the enemy's defensive positions may be bypassed using vertical envelopment from assault support assets. The commander may choose to conduct a double envelopment, and helicopterborne forces can be effectively used on a different route to attack than those of the GCE. This allows forces to converge with minimal risk of fratricide caused by two opposing friendly ground forces coming from different attack routes in a double envelopment. The ACE can screen the flanks of an enveloping force reducing its vulnerability to enemy counteraction.

Turning Movement. A turning movement may use aviation forces to pass around the enemy's principal defensive positions to secure by helicopterborne forces or fires objectives deep in the enemy's rear using the ACE's advantages in speed, range, and mobility. The turning force usually operates at such distances from the fixing forces that mutual support is unlikely, except in the case of aviation units that can mutually support ground forces because of speed, range, mobility, agility, and line of sight communications. The ACE can screen the flanks of a turning force reducing its vulnerability to enemy counteraction.

Infiltration. During infiltration, the ACE can—

- Achieve surprise.
- Occupy a position from which to support the main attack by fire, especially rotary-wing close air support assets that can hover or land.
- Conduct ambushes and raids in the enemy's rear area to harass and disrupt his command and control and support activities.
- Cut off enemy forward units.

However, without augmentation by the GCE, the ACE would have difficulty securing key terrain.

Penetration. A penetration is a form of offensive maneuver that seeks to breach the enemy's main defenses creating an assailable flank where none existed before. Aviation forces can create and support the penetration or they can attack the flanks once the break has been made through the enemy's main defenses.

Flanking Attack. Aviation forces work well when conducting a flanking attack because the enemy's strength is normally oriented to the front and aviation forces can use all of the battlespace to attack from the flanks to minimize the enemy's strengths.

Frontal Attack. Aviation forces are often used to create gaps with fires in the enemy's front or to prevent or delay enemy reinforcements reaching the frontlines. Normally, the ACE will support the GCE in a MAGTF frontal attack.

Ground Combat Element

The GCE is a task-organized, combined arms force that closes with and defeats the enemy through the use of fires and maneuver. The MAGTF greatly enhances the combined arms capabilities resident in the GCE by extending the battlespace through application of firepower, information operations, target acquisition, and mobility. The GCE is particularly effective in battlespace with restricted mobility, such as urban, wooded, mountainous or jungle. It is also highly effective in limited visibility and in missions to attack, defeat, and clear the enemy in prepared defenses.

To increase tactical tempo, flexibility, mobility, survivability, and to seize the initiative, as well as inflict shock effect on the enemy, the assault forces of the GCE can be transported by helicopter or organic assault amphibian vehicles. GCE mobility is often provided by a combination of these means.

Distribution of Forces

One of the primary ways the commander can influence the course of the attack is through the distribution of force into a main attack, one or more supporting attacks, and a reserve. By properly distributing his assets, the commander achieves superiority at the decisive time and place while maintaining the minimum necessary forces elsewhere to accomplish supporting tasks. The GCE's flexibility and capabilities are ideally suited for assignment to any of these missions.

Main Effort. The GCE commander provides the bulk of his combat power to the main effort to maintain momentum and ensure accomplishment of the mission. The commander personally allocates resources or shifts his main effort as needed. The GCE, together with other elements of the MAGTF, reconnoiters extensively to locate enemy strengths and weaknesses. Once a weakness is identified, the GCE commander rapidly maneuvers his main effort to exploit it.

The main effort is provided with the greatest mobility and the preponderance of combat support and combat service support. Consideration is made to the mobility, survivability, shock effect, sustainability, and lasting effect of the GCE when determining the force designated as the main effort. The commander normally gives the main effort priority of fire support.

Reserves are echeloned in depth to support exploitation of the main effort's success. The commander can further concentrate the main effort by assigning it a narrower zone of action. All other actions are designed to support the main effort.

Supporting Effort. The commander assigns the minimum combat power necessary to accomplish the purpose of each supporting effort. A supporting effort in the offense is carried out in conjunction with the main effort to achieve one or more of the following:

- Deceive the enemy as to the location of the main effort.
- Destroy or fix enemy forces that could shift to oppose the main effort.

- Control terrain that, if occupied by the enemy, will hinder the main effort.
- Force the enemy to commit reserves prematurely.

In support of the MAGTF single battle, the GCE can be an ideal supporting effort for the ACE when the ACE is assigned as the main effort. In logistic-oriented missions, such as humanitarian assistance operations, the GCE can be an ideal supporting effort for the combat service support detachment if that element is assigned as the main effort.

Reserves. The primary purpose of the reserve is to attack at the critical time and place to ensure the victory or exploit success. Its strength and location will vary with its contemplated mission, form of maneuver, terrain, possible enemy reaction, and clarity of the situation. The reserve should be:

- Positioned to readily reinforce the main effort.
- Employed to exploit success, not reinforce failure.
- Committed in strength, not piecemeal.
- Reconstituted immediately.

Types of Offensive Operations

An attack by the GCE rarely develops exactly as planned. The commander must be prepared to take advantage of fleeting opportunities that present themselves during offensive operations. To exploit these opportunities and generate tempo, command and control must be decentralized. Subordinate commanders must make decisions using their initiative and understanding of their senior's intent. In the attack, the GCE must minimize its exposure to enemy fire by using rapid maneuver and counterfire, exploiting cover offered by the terrain, avoiding obstacles, and maintaining security.

The GCE commander employs his organic fires and supporting arms in coordination with maneuver to enable him to close with the enemy. The commander prepares for the attack by successively delivering fires on enemy fire support assets, command and control assets, support facilities, and frontline units. These fires protect the force and restrict the enemy's ability to counter the attack. Artillery and other supporting arms ensure continuity of support and the ability to mass fires by timely displacement. During the final stages of the attack, the attacker must rely primarily on organic fires to overcome remaining enemy resistance.

The attack culminates in a powerful and violent assault. The assaulting units overrun the enemy using fire and movement. The attacker exploits success immediately by continuing to attack into the depth of the enemy to further disrupt his defense. Deep operations, augmented with ACE or other MAGTF fires and information operations, attack enemy command and control and critical logistic nodes or second echelon maneuver forces, helping to break down the enemy's cohesion. As the defense begins to disintegrate, the attacker pursues the enemy to defeat him completely.

Movement to Contact. Using its internal reconnaissance and security assets, in coordination with MAGTF and ACE capabilities, the GCE finds and maintains contact while developing the situation with ground combat enemy forces in order to achieve the commander's decisive action. The GCE commander will initiate contact with as minimal a force as necessary so as to maintain freedom of maneuver with the bulk of his force. Once contact is gained, it is not normally broken without authority from the MAGTF commander. The GCE commander must exercise careful judgment to ensure that by maintaining contact, his force is not bending to the will of the enemy or being drawn into an ambush or other consequential action.

Attack. In the MAGTF single battle, the ACE and GCE, when supported in depth by the combat service support element (CSSE), have complementary capabilities. When integrated for the purpose of the attack, these capabilities can significantly increase their combined effects on the enemy for greater tactical decisiveness. In an attack, the GCE commander prevents effective enemy maneuver or counteraction by seizing the initiative through the use of his organic intelligence and security elements while masking his true intentions. The GCE commander makes every effort to achieve surprise by such methods as attacking under cover of darkness or using terrain and/or weather to conceal his force as it closes with the enemy. Once the GCE has gained the advantage, the commander will focus his combat power against the enemy's center of gravity through its critical vulnerabilities in order to destroy it and exploit all advantages gained.

Exploitation. The GCE normally conducts an exploitation by continuing the attack with committed units or by launching an uncommitted unit into the attack through a passage of lines. The commander may commit his reserve as the exploitation force depending on the factors of METT-T. He will constitute a new reserve as soon as possible to defeat enemy counterattacks and to restore momentum to a stalled attack.

Pursuit. Success in the pursuit is particularly enhanced through extensive use of the ACE to support the GCE's rapid movement and to provide flank security. Combat service support planning by the GCE in advance of the initial attack must take into account success and ensure that the combat trains have the mobility to support an aggressive pursuit.

Forms of Maneuver

The GCE commander selects the best form of maneuver to support the MAGTF commander's concept of operation.

Envelopment. The most successful envelopments by the GCE require MAGTF resources and support from the ACE and CSSE. By nature, envelopments require surprise, superior mobility (ground and/or air) on the part of the enveloping force, the main effort, and success by the supporting efforts to fix the enemy in place.

Turning Movement. During a turning movement, the main effort usually operates at such a distance from supporting efforts that its units are beyond mutual supporting distance. Therefore, the GCE's main effort must be self-sufficient or integrated with highly mobile CSSEs in order to reach the objective before becoming decisively engaged. A turning movement is rarely executed by a GCE of less than division strength. Consideration should be made to use the ACE as a supporting effort to capitalize on its inherent mobility, speed, and range.

Flanking Attack. The GCE commander will use fires and terrain, and exploit weaknesses in enemy dispositions to create a flank. To the GCE, a flanking attack is similar to envelopment but is conducted on a shallower axis and is usually less decisive and less risky than a deeper attack. A flanking attack is usually conducted by battalions or below. This attack usually requires a supporting attack to occupy the enemy to the GCE's front.

Frontal Attack. The GCE goal in the frontal attack is to fix or defeat the enemy. The GCE commander may conduct feints or demonstrations in other areas to weaken the enemy effort at the breach by causing him to shift his reserves to the GCE's advantage.

Infiltration. The GCE commander must ensure that operational security is a top priority during planning and preparation for an infiltration as the forces conducting the infiltration are particularly vulnerable to surprise and ambush. Prearranged helicopter-delivered combat service support resupply is critical to support forces beyond the FEBA.

Penetration. The GCE must closely coordinate its operations with the ACE to take advantage of the ACE's ability to create gaps in the enemy's defense.

Combat Service Support Element

Combat service support planners should keep continuously informed of operation plans. They anticipate offensive operations even while supporting other types of operations. The objective of combat service support conducted in support of offensive operations is to extend operational reach and increase the endurance of the force by supporting as far forward as possible with a logistics system that is optimized for throughput.

To prepare for an attack, CSSEs ensure that all support equipment is ready and that supplies are best located for support. They ensure that enough transportation is available to support the tactical and support plans. Commanders ensure that all support elements understand their responsibilities.

The forward deployment of CSSEs must take into account the vulnerability of the unit to enemy counterattack and maneuver element requirements for space and roads. CSSEs, especially mobile combat service support detachments, require security assistance. They need to be written into the fire support plan, have their own list of on-call targets, and have assets to call for fire from artillery and aviation platforms, as well as have established procedures for actions upon enemy contact.

The fundamental principle of supply support in the offense is responsiveness-to the supported unit. Supply support is typically more difficult in the offense than in the defense because of the ever-changing locations of units and their support areas. The concept of support becomes even more important and increasingly difficult to execute. Combat service support planners must coordinate preparations and unit positioning with deception plans to avoid giving away the element of surprise. Consequently, most combat service support operations will be conducted under the cover of darkness.

Ammunition

Responsive ammunition support for offensive operations is critical. This support is more difficult in offensive operations due to the lengthening of supply lines and the need for user resupply vehicles to stay close to firing elements. In preparing for the attack, logistics planners consider the following:

- Placing ammunition close to the user.
- Preparing ammunition supply points and ammunition transfer points to rapidly move forward as the attack advances.
- Stockpiling artillery ammunition at designated firing positions (possibly forward of current positions).
- Moving ammunition forward with advancing elements to ensure that basic loads can be replenished quickly.

Fuel

Offensive operations use large quantities of fuel. As a result, logisticians prepare for the attack by building up stocks in forward sites while avoiding signaling intentions to the enemy. They also ensure that fuel supply elements can move forward as the attack develops. Control of bulk transporter assets must be closely maintained throughout the AO. This is particularly true if the attack is highly successful and results in exploitation or pursuit.

Maintenance

Planners ensure maintenance operations support momentum and massing at critical points. Maintenance personnel maximize momentum by repairing at the point of malfunction or damage. They enhance momentum by keeping the maximum number of weapon systems operable and mobile. Emphasis is on battle damage assessment and rapid return of equipment to the supported unit. Repair and recovery personnel perform their mission in forward areas.

Supply

While Classes III (petroleum, oils, and lubricants) and V (ammunition) are the most important supplies in the offense, planners consider all classes of supply. While the need for barrier and fortification material decreases, for example, the requirement for obstacle, breaching, and bridging material may increase. Weapons system requirements may also be higher since weapon systems exposure to enemy fire during offensive operations is usually greater.

Transportation and Distribution

Movement requirements heavily tax transportation resources. There may be a wide dispersion of units and lengthening lines of communications. There may also be an increased requirement for personnel replacements and some classes of supply, such as fuel and weapon systems. These factors demand close coordination and planning for the use of transportation assets. Techniques such as supply push (unit distribution) or mobile forward tactical resupply and refueling points may be incorporated into the concept of support. Resources such as transportation and supply infrastructure that may be secure in the more stable environment of defense may not be as reliable in the offense. The opening and securing of main supply routes and available logistics facilities to sustain the MAGTF's offensive operations must be included in the operational and combat service support planning.

The mobility of offensive operations requires reliance on motor and air transport. When considering the air transport mode, the planner also considers aerial delivery. Movement control personnel set priorities in accordance with the combatant commander's or joint force commander's priorities to ensure that transportation assets meet the most critical needs. Aerial delivery or external helicopter delivery may be in greater demand.

Medical

Offensive operations increase the burden on medical resources. Planners can expect high casualty rates. High casualties and long evacuation lines will stress medical treatment and evacuation resources to their limits and may dictate augmentation for medical detachments. Fleet hospitals move forward in preparation for offensive operations to provide maximum treatment and holding facilities. When organic medical resources are insufficient, evacuation may require use of nonmedical transportation assets, adding additional stress to an already overtaxed transportation system.

Services

The main combat service support effort in the offense is to provide only the most critically needed support to the attacking force. Most service functions play a minor role. Commanders suspend some services until the situation stabilizes. Laundry, clothing exchange, and field showers may be temporarily suspended. Mortuary affairs/graves registration is a major exception. It continues and may intensify. Adequate mortuary affairs/graves registration supplies must be on hand. Mortuary affairs detachments maintain close communications with personnel elements to verify and report casualty information and aid in the identification of remains.

DEFENSIVE OPERATIONS

An effective defense is never passive. The defender cannot prepare his positions and simply wait for the enemy to attack. Commanders at every level must seek every opportunity to wrest the initiative from the attacker and shift to the offense. Subordinate commanders take the necessary steps to maintain their positions and cover gaps in their dispositions by the use of observation, obstacles, fires or reserves. The defense demands resolute will on the part of all commanders.

Aviation Combat Element

The MAGTF commander uses speed, range, mobility, and agility of aviation assets to maximize concentration and flexibility in the defense. MAGTF aviation assets are integrated into MAGTF defensive operations either as the main effort or in a supporting role. During preparation for defensive operations, the ACE may support the covering force with aerial reconnaissance and fires. The

MAGTF commander may task-organize aviation, ground, and combat service support units under a single aviation combat commander to execute the form of defensive maneuver selected.

During defensive operations, the MAGTF commander organizes his battlespace into three areas: security area, main battle area, and rear area. The ACE will operate throughout all of these areas and is integral to the MAGTF's single battle in the defense.

Security Area

Typically, operations in the security area include interdiction by air maneuver and fires. During the defense, aviation can be used to attack deep against high-payoff targets, enemy concentrations, and moving columns, and also to disrupt enemy centers of gravity.

The MAGTF commander seeks to engage the enemy as far out as possible. Because of the mobility and range of aviation assets, the ACE has excellent capabilities to conduct these operations. ACE assets can be employed in depth to attack follow-on echelons before they can move forward to the main battle area. Aviation forces can be employed to conduct screening operations; in conjunction with ground forces, they conduct guard operations on an open flank. Normally, ACE forces are not given guard missions.

Main Battle Area

The greater the depth of the main battle area, the greater the maneuver space for maximizing the capabilities of the ACE. A counterattack is an attack by part or all of a defending force against an attacking enemy force, for such specific purposes as regaining ground lost or cutting off and destroying enemy advance units. ACE assets used as the counterattack force can be employed to conduct decisive action to regain the initiative.

Rear Area

MAGTF commanders should allocate adequate resources to protect the rear area to maintain freedom of action and continuity of operations. Aviation assets can support the force in the rear because of their range, speed, and mobility. Because ACE airfields often operate in rear areas, aviation assets must depend on those functions of security and sustainment required to maintain continuity of operations. Assault support assets increase the mobility of the tactical combat force that operates in the rear area. To increase reaction time, ACE assets may be employed as direct support assets to the rear area commander by the MAGTF commander.

Mobile Defense

Since minimum force is placed forward to canalize, delay, disrupt, and deceive the enemy as to the actual location of the defense, MAGTF aviation assets can supplement mobile forces to fill in gaps where the MAGTF is most vulnerable. A mobile defense requires mobility greater than that of the attacker. The MAGTF generates the mobility advantage with helicopterborne forces and MAGTF aviation assets. The ACE can support through fires the displacement of GCE units to alternate and supplementary positions used in the mobile defense. Terrain and space are traded to draw the enemy deeper into the defensive area, causing him to overextend his force and expose his flanks to ACE assets. Together, MAGTF aviation assets and ground combat forces provide a much more effective strike force that can bring simultaneous fires to bear upon the enemy from unexpected directions.

Position Defense

In a position defense, the MAGTF commander can employ his aviation assets (primarily assault support aircraft) to help contain tactical emergencies, by disengaging them from an area and quickly concentrating them in another. Because of the ACE's mobility and agility, the MAGTF commander can risk reducing the size of the ground maneuver force placed in reserve. In a position defense, aviation assets can be used to blunt and contain enemy penetrations, to counterattack, and to exploit opportunities presented by the enemy.

Ground Combat Element

The GCE conducts the defense through the assignment of sectors, battle or blocking positions, and strong points. These assignments are made in a manner that enhances depth and mutual support; provides opportunities to trap or ambush the attacker; and affords observation, surprise, and deception. The GCE commander maintains an awareness of concurrent delaying actions to take advantage of opportunities created by adjacent units. The GCE receives substantial heavy engineering and logistical support from the combat service support detachment to enhance the survivability, sustainability, and countermobility of its defensive positions. The ACE provides support to the GCE through assault support, close air support, and reconnaissance.

Security Forces

GCE security forces are employed in the security area to delay, disrupt, and provide early warning of the enemy's advance and to deceive him as to the true location of the main battle area. These forces are assigned cover, guard or screen missions.

Screening Force

The GCE may establish a screening force to gain and maintain contact with the enemy, observe enemy activity, identify the enemy main effort, and report information. In most situations, the minimum security force organized by the GCE is a screening force. Normally, the screening force only fights in self-defense, but may be tasked to—

- Repel enemy reconnaissance units as part of the GCE's counterreconnaissance effort.
- Prevent enemy artillery from acquiring terrain that enables frontline units to be engaged.
- Provide early warning.
- Attack the enemy with supporting arms.

Guard Force

The GCE may designate a guard force for protection from enemy ground observation, direct fire, and surprise attack for a given period of time. A guard force allows the commander to extend the defense in time and space to prevent interruption of the organization of the main battle area. Observation of the enemy and reporting of information by the guard force is an inherent task of the guard force, but secondary to its primary function of protection.

The GCE commander determines the orientation of the guard force and the duration the guard must be provided. Normally, guard forces are oriented to the flanks for the minimum amount of time necessary to develop an integrated defense.

Covering Force

The GCE may provide the bulk of the MAGTF's covering force. The covering force operates apart from the main force to engage, delay, disrupt, and deceive the enemy before he can attack the main force. A GCE covering force can be augmented or supported by rotary-wing attack assets in order to strengthen its capabilities and further disrupt enemy attack formations.

Security Measures

Security measures are employed by the GCE and coordinated at all levels. These security measures include combat patrolling, sensors, target acquisition radars, surveillance, and employment of false visual and electronic signatures. In addition, skills of certain units within the GCE enhance the security posture of the organization. For example, engineers within the GCE contribute to survivability, mobility, and countermobility, all of which contribute to security.

Any active measure that may impact on other elements of the MAGTF is coordinated throughout the MAGTF. All units of the GCE provide local security. The degree of local security is dictated by terrain, communications, target acquisition capabilities, and the enemy threat.

Combat Service Support Element

The role of the CSSE in the defense is to support defensive battles while maintaining the capability to shift to the offense with little notice. Facilities and combat service support areas should be far enough in the rear to be out of the flow of battle and relatively secure. They should not be so far back that they make the support effort less effective. Where possible, combat service support units locate out of the reach of potential penetrations in protected and concealed locations without sacrificing support and out of the movement routes for retrograding units. Dispersion should be consistent with support requirements, control, and local security. Air defense coverage should be planned and emplaced.

Ammunition

Logisticians position ammunition supply and transfer points to facilitate rapid and responsive support. Using units may stockpile ammunition in excess of their basic loads. Ammunition may also be placed at successive defensive positions. This provides easy access and lessens transportation problems during the withdrawal to those positions. The defense usually requires a greater volume of ammunition than the offense. Construction and barrier material and ammunition requirements, especially for mines and barrier materials, are heaviest during the preparation for defense.

Fuel

The form of defensive operation influences fuel requirements. A position defense typically requires less fuel than an offensive operation. Mobile defenses, on the other hand, generally involve greater fuel consumption that the more static-oriented area defense. In either case, forward stockpiles of fuel may be appropriate.

Maintenance

The primary thrust of the maintenance effort in the defense is to maximize the number of weapon systems available at the start of the operation. Once the defensive battle begins, the thrust is to fix the maximum number of inoperable systems and return them to battle in the least amount of time. This requires forward support at, or as near as possible to, the intended AO of the systems.

Supply

Supply activity will be the most intensive during the preparation stage. Stockpiles should be far forward and at successive defensive positions, especially critical supplies (fuel, ammunition, barrier materiel). While many supplies—especially munitions and barrier material—must be far forward, they must also be as mobile as possible. This allows continuous support as combat power shifts in response to enemy attacks. The CSSE must position the ammunition supply points or transfer points to maximize responsiveness.

Transportation and Distribution

Transportation resources are most critical in the preparation stage of the defense. Stockpiling supplies and shifting personnel, weapon systems, and supplies require extensive transportation, laterally or in depth, to meet the probable points of enemy attack. Transportation assets move barrier supplies and ammunition (e.g., mines, demolitions) as close to the barrier sites as possible. Logisticians take action to increase the flow of these materials as soon as the intention to conduct a deliberate defense is known.

Medical

Medical support of defensive operations is more difficult than in the offense. Casualty rates are lower, but forward acquisition is complicated by enemy action and the initial direction of maneuver to the rear. The task of frontline medical units is to stabilize, prioritize, and evacuate the wounded. Priorities for evacuation will be complicated by the probable enemy main effort. Enemy activities may inhibit evacuation, increase casualties among medical personnel, and damage medical and evacuation equipment. Heaviest casualties, including those caused by enemy artillery and weapons of mass destruction, may be expected during the initial enemy attack and in the counterattack.

The enemy attack may disrupt ground and air communications routes and delay evacuation of patients to and from aid stations. Clearing facilities should be located away from points of possible penetration and must not interfere with reserve force positioning. The depth and dispersion of the mobile defense create significant time and distance problems in evacuation support to security and fixing forces. Security forces may be forced to withdraw while simultaneously carrying their patients to the rear. Peak loads may require additional helicopter evacuation capability. Nonmedical transportation assets may not be available to assist in casualty evacuation.

Services

In the defense, services operate routinely where the tactical situation permits. Service facilities should locate out of the way and not interfere with tactical operations. Mortuary affairs detachments evacuate the dead as rapidly as possible especially in deliberate defensive position to maintain morale. The use of hot rations tends to increase in the defense. Aerial delivery of rations and other services may be employed for cut-off, screening or guarding units.

OTHER TACTICAL OPERATIONS

A MAGTF may be required to conduct other tactical operations in combination, sequentially or as part of the offense or defense. Such operations are difficult, complex, often involve risk, and require detailed planning. Methods for conducting other tactical operations vary according to METT-T factors as they apply to each situation.

Aviation Combat Element

The MAGTF commander uses the ACE's inherent capabilities of range, speed, mobility, and agility when conducting these tactical operations He should ensure that adequate battlespace is assigned to employ all the capabilities of available aviation assets. Marine aviation is capable of operating in any environment; however, weather can adversely affect its effectiveness in performing some functions such as assault support and reconnaissance. Longer periods of employment will require increased maintenance efforts and the MAGTF may be required to support the joint force commander by providing excess sorties.

Retrograde

Aviation plays a major role in setting the conditions for a successful retrograde. The ACE can provide security for friendly ground forces and interdict enemy forces to disrupt and delay his advance. Air delivered mines can be used to supplement obstacles emplaced by engineers to impede or canalize enemy movements throughout the battlespace. Assault support may be used to move ground forces rapidly between delaying positions and move troops, equipment, and supplies away from the enemy. When a retirement occurs over extended distances, the security mission may be given to the aviation commander and appropriate ground units may be placed under his command authority. Retrograde operations are conducted primarily during limited visibility; therefore, aviation's all-weather abilities should be exploited. Should the retrograde operation require the displacement of aviation assets, the MAGTF should plan for the movement by echelon of airfield equipment and personnel while maintaining continuous aviation support for the duration of the operation.

Passage of Lines

MAGTF aviation assets can support a forward passage of lines by providing or supporting the security force to fix enemy forces in place and permit the MAGTF to complete the passage of lines. Aviation could then be used to exploit success of the moving force. In a rearward passage, in addition to a security force role, aviation can serve as the MAGTF counterattack force.

Linkup

Tactical aviation assets can be used to establish initial electronic connectivity between two units conducting a linkup while physical contact between ground forces occurs later. A helicopterborne force acting as the moving force can usually accomplish physical linkup rapidly.

Relief in Place

Assault support assets are ideal to transport infantry units to conduct a rapid relief in place especially where there is no enemy pressure or where a replacement of like type units is required. In certain instances, a relief in place of a ground unit with an aviation force such as attack helicopters can keep the enemy off balance and rest a ground unit.

Obstacle Crossing

Aviation assets give the MAGTF the ability to cross obstacles with minimal delay, loss of momentum, and casualties. Helicopterborne forces can bypass most obstacles completely; if necessary, these forces can reduce or eliminate the obstacle from the far side of the impediment. Aviation assets can suppress and disrupt the enemy when the force is most vulnerable while astride the obstacle.

Breakout from Encirclement

Normally, when encircled by the enemy, a MAGTF commander will attempt to breakout as soon as possible. Aviation forces provide an immediately responsive and effective asset to aid in the breakout. The encircled force may receive fire support from aviation assets outside the encirclement. Attack helicopters can conduct a breakout by rupturing and penetrating the enemy's encircling position, widening the gap until all the other encircled forces have moved through. In addition, aviation forces can be used as a diversion or may augment the reserve when committed. Any of the encircled forces (rupture force, main body, or rear guard) may consist of aviation and ground task-organized combined arms teams.

Ground Combat Element

The GCE conducts other tactical operations to support the MAGTF's offensive and defensive operations. These operations may require augmentation of specialized equipment and personnel with special skills. The type of augmentation will depend on the characteristics of the AO, conditions under which they are conducted, the nature of the operations, or any combination of these factors. The GCE is dependent upon the rest of the MAGTF for the additional fires, logistics, and other support necessary to execute these operations with speed and security.

Retrograde

In a retrograde, the GCE will normally conduct disengagement by echelon. Security forces (such as the guard and covering force) and the reserve usually are highly mobile units comprised of tanks, light armored reconnaissance, and infantry mounted on assault amphibious vehicles and augmented by attack helicopter assets. The GCE's organic combat engineering assets or those from the CSSE are employed to prepare initial and subsequent delaying positions and support other countermobility requirements. Indirect fires are used to attack enemy formations, force their early deployment, slow their advance, and limit their contact with friendly forces. Tactical deception is used to confuse the enemy as to the true location and intent of ground forces; the retrograde itself may be a deception measure to make the enemy susceptible to a counterattack. The MAGTF commander should consider the use of a mobile reserve to support the counterattack during a retrograde. The GCE will employ appropriate force protection measures and normal movement-to-contact methods, including security measures, in a retrograde.

Passage of Lines

The GCE can control linkup operations between its subordinate commands or conduct them with the ACE, CSSE, and with other joint or multinational forces. When the GCE conducts a linkup, the force designated as the stationary unit should at least temporarily occupy the designated linkup point. The moving ground force commander will normally locate his forward command post in the vicinity of the stationary ground force combat operations center to facilitate integration and coordination of tactical plans, fire support, security, command and control, combat service support, communications, and maneuver control and fire support coordinating measures. The GCE commander should ensure that appropriate command relationships are established and understood by both elements. Fire support coordinating measures, such as restricted fire lines, are established or modified as required to balance freedom of action and positive control.

Linkup

When the linkup is between two subordinate ground units, the GCE establishes maneuver control measures such as linkup points and boundaries between converging forces, and fire support coordinating measures such as restricted fire lines and coordinated fire lines. Control measures are adjusted during the operation to provide for freedom of action and maximum control. The GCE commander may designate linkup points, usually located where the moving force's routes arrive at the location of the stationary force's security elements. Alternate linkup points are also designated since enemy action may interfere with linkup at primary points. To assist in the linkup, stationary forces help open lanes in minefields, breach or remove selected obstacles, furnish guides, and designate assembly areas. Leading elements of each force should be on a common radio net.

Relief in Place

Control of all ground units normally remains with the outgoing commander. This requires close coordination with the supported units. Units may need to exchange certain weapons, supplies, equipment, and, occasionally, vehicles to facilitate a rapid relief. To ensure coordination and maintain security, the outgoing unit's radio nets, command frequencies, and operators should be used. The outgoing unit remains in charge of communications throughout the entire relief. Artillery is normally relieved last to ensure continuous fire support; if possible, the outgoing unit artillery remains in position until all units are relieved.

Obstacle Crossing

The GCE normally bypasses obstacles whenever possible, often using helicopterborne forces conducting an envelopment. It has the capability to conduct hasty and deliberate breaches. When conditions permit, assault amphibious vehicles are ideal to move assault elements across a river. For large-scale river crossing operations, the GCE may require additional bridging assets provided by the MAGTF or the joint force. The GCE maximizes the use of combined arms during crossing operations. Use of supporting arms, combat engineers, reconnaissance, rotary wing ACE assets, and armor reduces vulnerability, increases tempo, and supports initiative in breaching operations. Deception is maximized to deceive the enemy and draw enemy attention away from the crossing site.

Breakout from Encirclement

The GCE will attempt to deceive the enemy on the time and place of the breakout. It will make best use of limited visibility but not necessarily at the expense of time. The GCE will use its organic reconnaissance as well as other reconnaissance assets to locate gaps and weaknesses in the enemy force. Initially, the rupture force will be the GCE's main effort and may be provided additional combat power, such as engineer support, necessary to achieve the rupture. During the breakout, massed continuous fires are used to open the rupture point, suppress enemy direct fire systems, and isolate the breakout from the enemy. Once the rupture is achieved, priority of fires may shift to the rear guard action if sufficient fires are available to support the momentum of the breakout. Artillery will provide continuous fire support during the breakout and subsequent movement to linkup with friendly forces.

Combat Service Support Element

The principles of logistics—responsiveness, simplicity, flexibility, economy, attainability, sustainability, and survivability—are universal constants that apply equally to the functional areas of logistics during other tactical operations. These considerations will not dictate a specific course of action, but will help maximize the effectiveness and efficiency of logistics operations.

Retrograde

Priority of support during retrograde operations is determined by the commander but is usually given to units that have completed the move and are preparing new positions. CSSEs must continue to support the delaying force with critical supplies at the old defensive positions while establishing support to withdrawing elements moving rearward. Combat service support personnel and equipment not essential to supporting forward combat forces should be moved as soon as feasible. Retrograde operations will strain the transportation system as all essential supplies, materiel, and personnel are moved rearward. Movement control personnel and agencies should maximize the use of all available transportation assets—watercraft, railroads, air, and line haul. All movements throughout the entire retrograde will be regulated, controlled, and prioritized to eliminate unnecessary surge periods and to avoid congestion. Helicopter and aerial delivery should be used whenever possible, as well as mobile loading of fuel and ammunition. If sufficient rolling stock is not available for mobile supply points, supplies can be placed along the retrograde route so forces can fall back on a continuous supply. Heavy equipment transportation should be coordinated by the senior movement control organization. Supplies that cannot be moved should be destroyed. Maintenance efforts should concentrate on use of controlled exchange and cannibalization to facilitate rapid turnaround of weapon systems. Repair to transportation assets is critical to retrograde operations.

Passage of Lines

The CSSE should establish liaison and coordinate movement control during the passage with the other force involved. Every action should be taken to avoid any interruption in logistics operations that would diminish the combat power of either force. All units should completely understand which unit will provide supply, maintenance; nuclear, biological, and chemical decontamination; medical; and movement priorities and control for the stationary and passing forces.

Linkup

Before the linkup has been initiated, combat service support is the responsibility of each unit involved in the linkup operation (whether it be converging forces, a force closing on a previous secured objective, forces encircling an enemy force, or during a counterattack). Converging forces should coordinate combat service support that can be mutually provided to facilitate the linkup operation and any subsequent mission.

Relief in Place

A CSSE involved in a relief in place should develop and coordinate a common concept of support, and exchange standing operating procedures, and combat support and combat service support status. The concept of support should clearly identify the specific elements of combat service support to be provided by each force involved in the relief. When possible, existing supplies, end items, and maintenance facilities should be left in place for the relieving force or prepositioned to support the movement of the forces involved in the relief.

Obstacle Crossing

Combat service support for obstacle crossing operations differs slightly from sustainment operations during the offense or defense. Transportation support for engineer units and bridging materiel is the primary concern, with maintenance of bridging equipment and fuel requirements a secondary consideration. All essential combat support and combat service support units should be moved across the obstacle early and dispersed in locations that can support the operation. Whenever possible, bridging equipment should be recovered early and replaced with assault float bridging and unit assets that can be recovered quickly. CSSEs may also have unique intelligence collection requirements, such as obstacle surveys or soil and trafficability studies, that must be satisfied in order to provide the desired support.

Breakout from Encirclement

The commander of an encircled force may have to reorganize his logistic support, centralize all supplies, and establish strict rationing and supply procedures to conserve his sustainment ability. If possible, resupply and casualty evacuation should be done by air. Centralized medical and graves registration operations should be established. The CSSEs should be integrated into the main body. In the event that some forces must be left behind, sufficient medical personnel and supplies will be left to attend the wounded, and personnel will also be detailed to destroy abandoned equipment.

APPENDIX E

Planning and Employment Considerations for MOOTW

Contents	
Unit Integrity.	E-1
Information Operations.	E-2
Civil-Military Coordination	E-3
Religious Ministry Support.	E-4
Legal.	E-4

Marine Corps forces may be required to conduct many of the various types of military operation other than war (MOOTW) while conducting combat operations in a major theater war. Plans for MOOTW are prepared in a similar manner as plans for war. As with combat operations, mission analysis and the commander's operational design are the foundation of the plan. The development of a clear definition, understanding, and appreciation for all potential threats is essential in assisting the commander in the proper organization of forces. All operations should be specific in nature with an established mission and end state. Rules of engagement should be clearly defined, fully disseminated, and reviewed for continued relevance as the situation or mission changes. Operations that are open-ended and not clearly defined can develop "mission creep," an insidious expansion of the original mission until there are either several new missions or the nature of the mission has changed. The commander and his staff should focus on the following key issues during planning.

UNIT INTEGRITY

To quickly deploy and begin operations immediately, planners should try to maintain unit integrity when developing a proposed force list for MOOTW. Marine forces train as units and are best able to accomplish a mission when deployed intact. By deploying as an existing unit, Marine forces are able to continue to operate under established procedures, adapt these procedures to the mission and situation, and maintain effectiveness. An ad hoc force is less effective and takes more time to adjust to the requirements of the mission. This not only complicates mission accomplishment, but may also have an impact on force protection. Even if political restraints on an operation dictate that a large force cannot be deployed intact, commanders should select smaller but cohesive elements that have trained and operated together. Additionally, when deploying into a situation involving combat operations, units should deploy with appropriate combat capability.

INFORMATION OPERATIONS

MOOTW require sound management of information to ensure unity of purpose and consistent themes in psychological, public affairs, and civil affairs operations. Psychological operations provide a planned, systematic process of conveying messages and influencing selected target groups. The messages are intended to promote particular themes that can result in desired attitudes and behaviors. This information may include safety, health, public service, and messages designed to favorably influence foreign perceptions of United States forces and operations. Success may hinge on direct control of or direct influence over the mass communication media (radio and television) in the region.

MOOTW often become of significant interest to the news media. As a result of this interest, the MAGTF commander must be prepared to deal with the media on a daily basis. Public affairs, including media reporting, influences public opinion and is a principal factor in the success or failure of MOOTW. Worldwide media coverage provided by satellite communications makes planning for public affairs more important than in the past. This is especially critical in MOOTW, where there can be significant political impact. Media reporting influences public opinion, which may affect the perceived legitimacy of an operation and ultimately influence the success or failure of the operation. The speed with which the media can collect and convey information to the public makes it possible for the world populace to become aware of an incident as quickly as, or even before, MAGTF decisionmakers.

MAGTFs should develop a well-defined, concise public affairs plan to minimize adverse effects upon the operation and integrate their public affairs officer early in the planning process. The MAGTF should facilitate open and independent reporting, respond to media queries that provide the maximum disclosure with minimum delay, and create an environment between commander and reporters that encourages balanced coverage of operations. An effective public affairs plan provides ways to communicate information about an operation and fulfills the military's obligation to keep the American public informed. It enhances force protection by ensuring that the media does not compromise operational security. The public affairs plan promotes operational security awareness by addressing the possibility of media attempts to acquire and publicly disseminate classified information. Public affairs plans should also anticipate responses to inaccurate media analysis and promulgation of disinformation and misinformation.

The command must establish a positive relationship with host-nation leadership and the local populace. Civil affairs units ensure local needs are identified and build cooperation and coordination among the participants. They assess the civil infrastructure, assist in the operation of temporary shelters, and serve as liaison between military and outside groups. These units contain a variety of specialty skills that support small wars and contingencies and are normally tailored to support specific operational requirements. They can also provide expertise on factors that directly affect military operations to include culture, social structure, economic systems, language, and host-nation support capabilities. Civil affairs units may also include forces conducting activities normally the responsibility of local or indigenous governments. Selection of civil affairs units should be based on a clear concept of the mission requirements for the type operation being planned.

CIVIL-MILITARY COORDINATION

MOOTW are normally joint and multinational operations set in an interagency environment. In many cases, nongovernmental agencies, media concerns, and other nontraditional influences will affect decisionmaking. Coordination with nongovernmental organizations, international organizations, and interagency operations allows the MAGTF to gain greater situational and cultural awareness. A technique to build unity of effort and conduct liaison with nonmilitary organizations is the establishment of a civil-military operations center (CMOC). Members of a CMOC may include representatives of adjacent and allied military commands, U.S. government agencies, other countries' forces involved in the operation, and civilian organizations. Civil affairs units should be the core of the CMOC. Through a CMOC, the MAGTF can gain a greater understanding of the roles of civilian organizations and how they influence mission accomplishment. Although formal agreements are not always necessary, such agreements between military and civilian organizations may improve coordination and effectiveness.

MOOTW can involve other United States non-Department of Defense departments and agencies. Within the United States, the Federal Emergency Management Agency normally leads the response to a natural disaster, while the Departments of Justice or Transportation could be expected to lead in a counterterrorist operation. Effective liaison with the lead agency enables the

MAGTF to support the political objectives of the operation. Outside the United States, the lead agency will normally be the Department of State and the United States ambassador will coordinate activities through an established country team with representation from all United States departments and agencies in that country. A non-Department of Defense lead agency does not alter the military chain of command.

RELIGIOUS MINISTRY SUPPORT

The cultural complexity of certain MOOTW elicits the need for special consideration regarding religious ministry. The role of religious ministry teams may be expanded beyond the traditional provision of divine services and facilitation of religious requirements within the unit. As advisors to the commander, chaplains assess and advise commanders and their staffs on cultural and religious issues—as well as moral, ethical, morale, and core values—both internal and external to the command. This includes advising the commanders and their staffs on religious and cultural matters within the AO, on ethical issues impacting the mission, and on morale within the indigenous community. Additionally chaplains play a major role in support of noncombatant evacuation operations, foreign humanitarian assistance/ disaster relief operations, or peacekeeping operations and should be active members of the CMOC. Religious ministry teams should be trained adequately and provided the resources and command support necessary to perform these functions. Likewise, commanders and their staffs should be trained to understand the role chaplains should play in the planning process as well as the input they have to the commander's situational awareness.

LEGAL

MOOTW may present unique legal challenges. In addition to traditional skills necessary in military justice, legal personnel may require expertise in areas such as refugees, displaced and detained civilians, fiscal law, rules of engagement, psychological operations, civil affairs, local culture, customs, and government, international, and environmental law. Commanders should ensure that the staff judge advocate has the resources available to respond to the variety of complex international and operational legal and regulatory issues that may arise. Host-nation legal personnel can be integrated into the MAGTF legal staff as soon as practical to provide guidance on unique indigenous legal practices and customs. The MAGTF commander should be alert for potential legal problems arising from the unique, difficult circumstances and the highly political nature of such operations as disaster relief and humanitarian assistance.

One of the United States military's most experienced leaders in the field of MOOTW, General Anthony Zinni, USMC (Retired), has developed the following considerations for humanitarian assistance, peacekeeping, and peace enforcement operations:

- Each operation is unique. We must be careful what lessons we learn from a single experience.
- Each operation has two key aspects: (1) the degree of complexity of the operation, and (2) the degree of consent of the involved parties and the international community for the operation.
- The earlier the involvement, the better the chance for success.
- Start planning as early as possible, including everyone in the planning process.
- Make as thorough an assessment as possible before deployment.
- Conduct a thorough mission analysis, determining the centers of gravity, end state, commander's intent, measures of effectiveness, exit strategy, and the estimated duration of the operation.
- Stay focused on the mission. Line up military tasks with political objectives. Avoid mission creep and allow for mission shifts. A mission shift is a conscious decision, made by the political leadership in consultation with the military commander, responding to a changing situation.
- Centralize planning and decentralize execution of the operation. This allows subordinate commanders to make appropriate adjustments to meet their individual situation or rapidly changing conditions.
- Coordinate everything with everybody. Establish coordination mechanisms that include political, military, nongovernmental organizations, international organizations, and the interested parties.
- Know the culture and the issues. We must know who the decisionmakers are. We must know how the involved parties think. We cannot impose our cultural values on people with their own culture.
- Start or restore key institutions as early as possible.
- Don't lose the initiative and momentum.
- Don't make unnecessary enemies. If you do, don't treat them gently. Avoid mindsets or use words that might come back to haunt you.
- Seek unity of effort and unity of command. Create the fewest possible seams between organizations and involved parties.
- Open a dialogue with everyone. Establish a forum for each of the involved parties.
- Encourage innovation and nontraditional responses.
- Personalities often are more important than processes. You need the right people in the right places.
- Be careful whom you empower. Think carefully about who you invite to participate, use as a go-between, or enter into contracts with since you are giving them influence in the process.
- Decide on the image you want to portray and keep focused on it. Whatever the image, humanitarian or as firm but well-intentioned agent of change, ensure your troops are aware of it so they can conduct themselves accordingly.
- Centralize information management. Ensure that your public affairs and psychological operations are coordinated, accurate, and consistent.
- Seek compatibility in all operations; cultural and political compatibility and military interoperability are crucial to success. The interests, cultures, capabilities, and motivations of all the parties may not be uniform, but they cannot be allowed to work against each other.
- Senior commanders and their staffs need the most education and training in nontraditional roles. The troops need awareness and understanding of their roles. The commander and the staff need to develop and apply new skills, such as negotiating, supporting humanitarian organizations effectively and appropriately, and building coordinating agencies with humanitarian goals.

APPENDIX F

Glossary

Section I Acronyms

ACE	aviation combat element
ACF	air contingency force
ADCON	administrative control
AO	area of operations
AOI	area of interest
AT	antiterrorism
ATF	amphibious task force
BLT	battalion landing team
CAP	combined action platoon
CBAE	commander's battlespace area evaluation
CCIR	commander's critical information requirements
CE	command element
CHOP	change of operational control
CI	counterintelligence
CINCs	commanders in chief
CINCUSJFCOM	Commander in Chief, U.S. Joint Forces Command
CINCUSPACOM	Commander in Chief, U.S. Pacific Command
CMOC	civil-military operations center
COA	course of action
COCOM	combatant command (command authority)
COMCMFC	Commander, Combined Marine Forces Command
COMMARCENT	Commander, Marine Corps Forces, Central
COMMARFOREUR	Commander, Marine Corps Forces, Europe
COMMARFORLANT	Commander, Marine Corps Forces, Atlantic
COMMARFORPAC	Commander, Marine Corps Forces, Pacific
COMMARFORSOUTH	Commander, Marine Corps Forces, South
COMUSMARFOR-K	Commander, U.S. Marine Corps Forces-Korea
CONUS	continental United States
CSS	combat service support
CSSE	combat service support element
DOD	Department of Defense
DODD	Department of Defense directive
EMW	expeditionary maneuver warfare

FAST	Fleet Antiterrorism Security Team
FEBA	forward edge of the battle area
FROG	free rocket over ground
FSCL	fire support coordination line
GCE	ground combat element
HUMINT	human intelligence
IO	information operations
JP	joint publication
JSTARS	Joint Surveillance Target Attack Radar System
LF	landing force
LOGAIS	logistics automated information system
MAG	Marine aircraft group
MAGTF	Marine air-ground task force
MALS	Marine aviation logistics squadron
MARCORSYSCOM	Marine Corps Systems Command
MARFORCENT	Marine Corps Forces, Central
MARFOREUR	Marine Corps Forces, Europe
MARFORLANT	Marine Corps Forces, Atlantic
MARFORPAC	Marine Corps Forces, Pacific
MARFORSOUTH	Marine Corps Forces, South
MCCDC	Marine Corps Combat Development Command
MCDP	Marine Corps doctrinal publication
MCMC	Marine Corps Materiel Command
MCPP	Marine Corps Planning Process
MCRP	Marine Corps reference publication
MCWP	Marine Corps warfighting publication
MEB	Marine expeditionary brigade
MEF	Marine expeditionary force
METT-T	mission, enemy, terrain and weather, troops and support available - time available
MEU	Marine expeditionary unit
MEU(SOC)	Marine expeditionary unit (special operations capable)
MLC	Marine Logistics Command
MOOSEMUSS	mass; objective; offensive; security; economy of force; maneuver; unity of command; surprise; simplicity
MOOTW	military operations other than war
MPF	maritime pre-positioning force
MPS	maritime pre-positioning ship
MPSRON	maritime pre-positioning ships squadron
NEF	naval expeditionary forces
OMFTS	operational maneuver from the sea

OPCON	operational control
OPS	operations
RLT	regimental landing team
RSOI	reception, staging, onward movement, and integration
SAR	synthetic aperture radar
SEALS	sea-air-land teams
SPMAGTF	special purpose MAGTF
STOM	ship-to-objective maneuver
TACON	tactical control
UAV	unmanned aerial vehicle
UNAAF	Unified Action Armed Forces
U.S.	United States
USCINCCENT	Commander in Chief, U.S. Central Command
USCINCEUR	Commander in Chief, U.S. European Command
USCINCJFCOM	Commander in Chief, U.S. Joint Forces Command
USCINCPAC	Commander in Chief, U.S. Pacific Command
USCINCSO	Commander in Chief, U.S. Southern Command
USMC	United States Marine Corps
USSOCOM	U.S. Southern Command
USTRANSCOM	United States Transportation Command

Section II Definitions

administrative control—Direction or exercise of authority over subordinate or other organizations in respect to administration and support, including organization of Service forces, control of resources and equipment, personnel management, unit logistics, individual and unit training, readiness, mobilization, demobilization, discipline, and other matters not included in the operational missions of the subordinate or other organizations. Also called ADCON. (JP 1-02)

amphibious assault—The principal type of amphibious operation that involves establishing a force on a hostile or potentially hostile shore. (JP 1-02)

amphibious demonstration—A type of amphibious operation conducted for the purpose of deceiving the enemy by a show of force with the expectation of deluding the enemy into a course of action unfavorable to him. (JP 1-02)

amphibious force—An amphibious task force and a landing force together with other forces that are trained, organized, and equipped for amphibious operations. (JP 3-02)

amphibious raid—A type of amphibious operation involving swift incursion into or temporary occupation of an objective followed by a planned withdrawal. (JP 1-02)

amphibious task force—A Navy task organization formed to conduct amphibious operations. (JP 3-02)

amphibious withdrawal—A type of amphibious operation involving the extraction of forces by sea in naval ships or craft from a hostile or potentially hostile shore. (JP 1-02)

area of influence—A geographical area wherein a commander is directly capable of influencing operations by maneuver or fire support systems normally under the commander's command or control. Also called AOI. (JP 1-02)

area of interest—That area of concern to the commander, including the area of influence, areas adjacent thereto, and extending into enemy territory to the objectives of current or planned operations. This area also includes areas occupied by enemy forces who could jeopardize the accomplishment of the mission. Also called AOI. (JP 1-02)

area of operations—An operational area defined by the joint force commander for land and naval forces. Areas of operation do not typically encompass the entire operational area of the joint force commander, but should be large enough for component commanders to accomplish their missions and protect their forces. Also called AO. (JP 1-02)

asymmetry—Unconventional, unexpected, innovative or disproportional means used to gain advantage over an adversary. (MCRP 5-12C)

attack—An offensive action characterized by movement supported by fire with the objective of defeating or destroying the enemy. (MCRP 5-12C)

aviation combat element—The core element of a Marine air-ground task force (MAGTF) that is task-organized to conduct aviation operations. The aviation combat element (ACE) provides all or a portion of the six functions of Marine aviation necessary to accomplish the MAGTF's mission. These functions are antiair warfare, offensive air support, assault support, electronic warfare, air reconnaissance, and control of aircraft and missiles. The ACE is usually composed of an aviation unit headquarters and various other aviation units or their detachments. It can vary in size from a small aviation detachment of specifically required aircraft to one or more Marine aircraft wings. The ACE itself is not a formal command. Also called ACE. (JP 1-02)

avenue of approach—An air or ground route of an attacking force of a given size leading to its objective or to key terrain in its path. Also called AA.(JP 1-02)

axis of advance—A line of advance assigned for purposes of control; often a road or a group of roads, or a designated series of locations, extending in the direction of the enemy. (JP 1-02)

barrier—A coordinated series of obstacles designed or employed to channel, direct, restrict, delay, or stop the movement of an opposing force and to impose additional losses in personnel, time, and equipment on the opposing force. Barriers can exist naturally, be manmade, or a combination of both. (JP 1-02)

battalion landing team—In an amphibious operation, an infantry battalion normally reinforced by necessary combat and service elements; the basic unit for planning an assault landing. Also called BLT.

battle position—1. In ground operations, a defensive location oriented on an enemy avenue of approach from which a unit may defend. 2. In air operations, an airspace coordination area containing firing points for attack helicopters. Also called BP. (MCRP 5-12C)

battlespace—1. The environment, factors, and conditions that must be understood to successfully apply combat power, protect the force, or complete the mission.

This includes the air, land, sea, space, and the included enemy and friendly forces; facilities; weather; terrain; the electromatic spectrum; and the information environment within the operational areas and areas of interest. (JP 1-02) 2. All aspects of air, surface, subsurface, land, space, and electromagnetic spectrum which encompass the area of influence and area of interest. (MCRP 5-12C)

battlespace dominance—The degree of control over the dimensions of the battlespace which enhances friendly freedom of action and denies enemy freedom of action. It permits force sustainment and application of power projection to accomplish the full range of potential operational and tactical missions. It includes all actions conducted against enemy capabilities to influence future operations. (MCRP 5-12C)

boundary—A line that delineates surface areas for the purpose of facilitating coordination and deconfliction of operations between adjacent units, formations, or areas. (JP 1-02)

breach—The employment of any means available to break through or secure a passage through an obstacle. (MCRP 5-12C)

centers of gravity—Those characteristics, capabilities, or localities from which a military force derives its freedom of action, physical strength, or will to fight. Also called COGs. (JP 1-02)

close operations—Military actions conducted to project power decisively against enemy forces which pose an immediate or near term threat to the success of current battles or engagements. These military actions are conducted by committed forces and their readily available tactical reserves, using maneuver and combined arms. (MCRP 5-12C)

combatant command (command authority)—Nontransferable command authority established by title 10 ("Armed Forces"), United States Code, section 164, exercised only by commanders of unified or specified combatant commands unless otherwise directed by the President or the Secretary of Defense. Combatant command (command authority) cannot be delegated and is the authority of a combatant commander to perform those functions of command over assigned forces involving organizing and employing commands and forces, assigning tasks, designating objectives, and giving authoritative direction over all aspects of military operations, joint training, and logistics necessary to accomplish the missions assigned to the command. Combatant command (command authority) should be exercised through the commanders of subordinate organizations. Normally this authority is exercised through subordinate joint force commanders and Service and/or functional component commanders. Combatant command

(command authority) provides full authority to organize and employ commands and forces as the combatant commander considers necessary to accomplish assigned missions. Operational control is inherent in combatant command (command authority). Also called COCOM. (JP 1-02)

combatant commander—A commander in chief of one of the unified or specified combatant commands established by the President. Also called CINC. (JP 1-02)

combat power—The total means of destructive and/or disruptive force which a military unit/formation can apply against the opponent at a given time. (JP 1-02)

combat service support—The essential capabilities, functions, activities, and tasks necessary to sustain all elements of operating forces in theater at all levels of war. Within the national and theater logistic systems, it includes but is not limited to that support rendered by service forces in ensuring the aspects of supply, maintenance, transportation, health services, and other services required by aviation and ground combat troops to permit those units to accomplish their missions in combat. Combat service support encompasses those activities at all levels of war that produce sustainment to all operating forces on the battlefield. Also called CSS. (JP 1-02)

combat service support area—An area ashore that is organized to contain the necessary supplies, equipment, installations, and elements to provide the landing force with combat service support throughout the operation. Also called CSSA. (JP 1-02)

combat service support element—The core element of a Marine air-ground task force (MAGTF) that is task-organized to provide the combat service support necessary to accomplish the MAGTF mission. The combat service support element varies in size from a small detachment to one or more force service support groups. It provides supply, maintenance, transportation, general engineering, health services, and a variety of other services to the MAGTF. The combat service support element itself is not a formal command. Also called CSSE. (JP 1-02)

combined arms—The full integration of combat arms in such a way that to counteract one, the enemy must become more vulnerable to another. (MCRP 5-12C)

command and control—The exercise of authority and direction by a properly designated commander over assigned and attached forces in the accomplishment of the mission. Command and control functions are performed through an arrangement of personnel, equipment, communications, facilities, and procedures employed by a commander in planning, directing, coordinating, and controlling forces and operations in the accomplishment of the mission. Also called C2. (JP 1-02)

command element—The core element of a Marine air-ground task force (MAGTF) that is the headquarters. The command element is composed of the commander, general or executive and special staff sections, headquarters section, and requisite communications support, intelligence, and reconnaissance forces necessary to accomplish the MAGTF mission. The command element provides command and control, intelligence, and other support essential for effective planning and execution of operations by the other elements of the MAGTF. The command element varies in size and composition. Also called CE. (JP 1-02)

commander's intent—A commander's clear, concise articulation of the purpose(s) behind one or more tasks assigned to a subordinate. It is one of two parts of every mission statement which guides the exercise of initiative in the absence of instructions. (MCRP 5-12C)

component—1. One of the subordinate organizations that constitute a joint force. Normally a joint force is organized with a combination of Service and functional components. (JP 1-02)

covering force—1. A force operating apart from the main force for the purpose of intercepting, engaging, delaying, disorganizing, and deceiving the enemy before the enemy can attack the force covered. 2. Any body or detachment of troops which provides security for a larger force by observation, reconnaissance, attack, or defense, or by any combination of these methods. (JP 1-02)

critical capability—An inherent ability that enables a center of gravity to function as such. Also called CC. (MCRP 5-12C)

critical requirement—An essential condition, resource, or means that is needed for a critical capability to be fully functional. Also called CR.(MCRP 5-12C)

critical vulnerability—An aspect of a center of gravity that if exploited will do the most significant damage to an adversary's ability to resist. A vulnerability cannot be critical unless it undermines a key strength. Also called CV. (MCRP 5-12C)

culminating point—The point in time and space when the attacker can no longer accomplish his purpose, or when the defender no longer has the ability to accomplish his purpose. This can be due to factors such as combat power remaining, logistic support, weather, morale, and fatigue. (MCRP 5-12A)

deception—Those measures designed to mislead the enemy by manipulation, distortion, or falsification of evidence to induce the enemy to react in a manner prejudicial to the enemy's interests. (JP 1-02)

deep operations—Military actions conducted against enemy capabilities which pose a potential threat to friendly forces. These military actions are designed to isolate, shape, and dominate the battlespace and influence future operations. (MCRP 5-12C)

defense—A coordinated effort by a force to defeat an attack by an opposing force and prevent it from achieving its objectives. (MCRP 5-12C)

defense in depth—The siting of mutually supporting defense positions designed to absorb and progressively weaken attack, prevent initial observations of the whole position by the enemy, and to allow the commander to maneuver the reserve. (JP 1-02)

defensive operations—Operations conducted with the immediate purpose of causing an enemy attack to fail. Defensive operations also may achieve one or more of the following: gain time; concentrate forces elsewhere; wear down enemy forces as a prelude to offensive operations; and retain tactical, strategic, or political objectives. (MCRP 5-12C)

delaying operation—An operation in which a force under pressure trades space for time by slowing down the enemy's momentum and inflicting maximum damage on the enemy without, in principle, becoming decisively engaged. (JP 1-02)

deliberate breaching—The creation of a lane through a minefield or a clear route through a barrier or fortification, which is systematically planned and carried out. (JP 1-02)

demonstration—1. An attack or show of force on a front where a decision is not sought, made with the aim of deceiving the enemy. 2. In military deception, a show of force in an area where a decision is not sought, made to deceive an adversary. It is similar to a feint, but no actual contact with the adversary is intended. (JP 1-02)

deterrence—The prevention from action by fear of the consequences. Deterrence is a state of mind brought about by the existence of a credible threat of unacceptable counteraction. (JP 1-02)

encircling force—In pursuit operations, the force which maneuvers to the rear or flank of the enemy to block its escape so that it can be destroyed between the direct pressure and encircling force. This force advances or flies along routes paralleling the enemy's line of retreat. If the encircling force cannot outdistance the enemy to cut it off, the encircling force may attack the enemy's flanks. (MCRP 5-12C)

end state—What the National Command Authorities want the situation to be when operations conclude-both military operations, as well as those where the military is in support of other instruments of national power. (JP 1-02)

feint—A limited-objective attack involving contact with the enemy, varying in size from a raid to a supporting attack. Feints are used to cause the enemy to react in three predictable ways: to employ reserves improperly, to shift supporting fires, or to reveal defensive fires. (MCRP 5-12C)

force protection—Actions taken to prevent or mitigate hostile actions against Department of Defense personnel (to include family members), resources, facilities, and critical information. These actions conserve the force's fighting potential so it can be applied at the decisive time and place and incorporates the coordinated and synchronized offensive and defensive measures to enable the effective employment of the joint force while degrading opportunities for the enemy. Force protection does not include actions to defeat the enemy or protect against accidents, weather, or disease. (JP 1-02)

forcible entry—Seizing and holding of a military lodgment in the face of armed opposition. (JP 3-18)

forward deployment—A basic undertaking which entails stationing of alert forces with their basic stocks for extended periods of time at either land-based overseas facilities or, in maritime operations, aboard ships at sea as a means of enhancing national contingency response capabilities. (MCRP 5-12C)

forward edge of the battle area—The foremost limits of a series of areas in which ground combat units are deployed, excluding the areas in which the covering or screening forces are operating, designed to coordinate fire support, the positioning of forces, or the maneuver of units. Also called FEBA. (JP 1-02)

frontal attack—An offensive maneuver in which the main action is directed against the front of the enemy forces. (JP 1-02)

functional component command—A command normally, but not necessarily, composed of forces of two or more Military Departments which may be established across the range of military operations to perform particular operational missions that may be of short duration or may extend over a period of time. (JP 1-02)

ground combat element—The core element of a Marine air-ground task force (MAGTF) that is task-organized to conduct ground operations. It is usually constructed around an infantry organization but can vary in size from a small ground unit of any type, to one or more Marine divisions that can be independently maneuvered under the direction of the MAGTF commander. The ground combat element itself is not a formal command. Also called GCE. (JP 1-02)

guard—A form of security operation whose primary task is to protect the main force by fighting to gain time while also observing and reporting information. (excerpt from JP 1-02)

humanitarian assistance—Programs conducted to relieve or reduce the results of natural or manmade disasters or other endemic conditions such as human pain, disease, hunger, or privation that might present a serious threat to life or that can result in great damage to or loss of property. Humanitarian assistance provided by US forces is limited in scope and duration. The assistance provided is designed to supplement or complement the efforts of the host nation civil authorities or agencies that may have the primary responsibility for providing humanitarian assistance. Also called HA. (JP 1-02)

joint force air component commander—The joint force air component commander derives authority from the joint force commander who has the authority to exercise operational control, assign missions, direct coordination among subordinate commanders, redirect and organize forces to ensure unity of effort in the accomplishment of the overall mission. The joint force commander will normally designate a joint force air component commander. The joint force air component commander's responsibilities will be assigned by the joint force commander (normally these would include, but not be limited to, planning, coordination, allocation, and tasking based on the joint force commander's apportionment decision). Using the joint force commander's guidance and authority, and in coordination with other Service component commanders and other assigned or supporting commanders, the joint force air component commander will recommend to the joint force commander apportionment of air sorties to various missions or geographic areas. Also called JFACC. (JP 1-02)

joint force commander—A general term applied to a combatant commander, subunified commander, or joint task force commander authorized to exercise combatant command (command authority) or operational control over a joint force. Also called JFC. (JP 1-02)

joint force land component commander—The commander within a unified command, subordinate unified command, or joint task force responsible to the establishing commander for making recommendations on the proper employment of land forces, planning and coordinating land operations, or accomplishing such operational missions as may be assigned. The joint force land component commander is given the authority necessary to accomplish missions and tasks assigned by the establishing commander. The joint force land component commander will normally be the commander with the preponderance of land forces and the requisite command and control capabilities. Also called JFLCC. (JP 1-02)

joint force maritime component commander—The commander within a unified command, subordinate unified command, or joint task force responsible to the establishing commander for making recommendations on the proper employment of maritime forces and assets, planning and coordinating maritime operations, or accomplishing such operational missions as may be assigned. The joint force maritime component commander is given the authority necessary to accomplish missions and tasks assigned by the establishing commander. The joint force maritime component commander will normally be the commander with the preponderance of maritime forces and the requisite command and control capabilities. Also called JFMCC. (JP 1-02)

joint logistics—The art and science of planning and carrying out, by a joint force commander and staff, logistic operations to support the protection, movement, maneuver, firepower, and sustainment of operating forces of two or more Military Departments of the same nation. (JP 1-02)

joint operations area—An area of land, sea, and airspace, defined by a geographic combatant commander or subordinate unified commander, in which a joint force commander (normally a joint task force commander) conducts military operations to accomplish a specific mission. Joint operations areas are particularly useful when operations are limited in scope and geographic area or when operations are to be conducted on the boundaries between theaters. Also called JOA. (JP 1-02)

joint task force—A joint force that is constituted and so designated by the Secretary of Defense, a combatant commander, a subunified commander, or an existing joint task force commander. Also called JTF. (JP 1-02)

landing force—A Marine Corps or Army task organization formed to conduct amphibious operations. (JP 3-02)

limit of advance—An easily recognized terrain feature beyond which attacking elements will not advance. (MCRP 5-12C)

line of communications—A route, either land, water, and/or air, that connects an operating military force with a base of operations and along which supplies and military forces move. Also called LOC. (JP 1-02)

linkup—An operation wherein two friendly ground forces join together in a hostile area. (MCRP 5-12C)

main body—The principal part of a tactical command or formation. It does not include detached elements of the command such as advance guards, flank guards, covering forces, etc. (MCRP 5-12C)

main effort—The designated subordinate unit whose mission at a given point in time is most critical to overall mission success. It is usually weighted with the preponderance of combat power and is directed against a center of gravity through a critical vulnerability. (MCRP 5-12C)

maneuver warfare—A warfighting philosophy that seeks to shatter the enemy's cohesion through a variety of rapid, focused, and unexpected actions which create a turbulent and rapidly deteriorating situation with which the enemy cannot cope. (MCRP 5-12C)

Marine air-ground task force—The Marine Corps principal organization for all missions cross the range of military operations, composed of forces task-organized under a single commander capable of responding rapidly to a contingency anywhere in the world. The types of forces in the Marine air-ground task force (MAGTF) are functionally grouped into four core elements: a command element, an aviation combat element, a ground combat element, and a combat service support element. The four core elements are categories of forces, not formal commands. The basic structure of the MAGTF never varies, though the number, size, and type of Marine Corps units comprising each of its four elements will always be mission dependent. The flexibility of the organizational structure allows for one or more subordinate MAGTFs to be assigned. Also called MAGTF. (JP 1-02)

Marine expeditionary brigade—A Marine air-ground task force that is constructed around a reinforced infantry regiment, a composite Marine aircraft group, and a brigade service support group. The Marine expeditionary brigade (MEB), commanded by a general officer, is task-organized to meet the requirements of a specific situation. It can function as part of a joint task force, or as the lead echelon of the Marine expeditionary force (MEF), or alone. It varies in size and composition, and is larger than a Marine expeditionary unit but smaller than a MEF. The MEB is capable of conducting missions across the full range of military operations. It may contain other Service or foreign military forces assigned or attached. Also called MEB. (proposed for JP 1-02)

Marine expeditionary force—The largest Marine air-ground task force (MAGTF) and the Marine Corps principal warfighting organization, particularly for larger crises or contingencies. It is task-organized around a permanent command element and normally contains one or more Marine divisions, Marine aircraft wings, and Marine force service support groups. The Marine expeditionary force is capable of missions across the range of military operations, including amphibious assault and sustained operations ashore in any environment. It can operate from a sea base, a land base, or both. Also called MEF. (JP 1-02)

Marine expeditionary unit—A Marine air-ground task force (MAGTF) that is constructed around an infantry battalion reinforced, a helicopter squadron reinforced, and a task-organized combat service support element. It normally fulfills Marine Corps forward sea-based deployment requirements. The Marine expeditionary unit provides an immediate reaction capability for crisis response and is capable of limited combat operations. Also called MEU. (JP 1-02)

Marine expeditionary unit (special operations capable)—The Marine Corps standard, forward-deployed, sea-based expeditionary organization. The Marine expeditionary unit (special operations capable) (MEU[SOC]) is a Marine expeditionary unit, augmented with selected personnel and equipment, that is trained and equipped with an enhanced capability to conduct amphibious operations and a variety of specialized missions of limited scope and duration. These capabilities include specialized demolition, clandestine reconnaissance and surveillance, raids, in-extremis hostage recovery, and enabling operations for follow-on forces. The MEU(SOC) is not a special operations force but, when directed by the National Command Authorities, the combatant commander, and/or other operational commander, may conduct limited special operations in extremis, when other forces are inappropriate or unavailable. Also called MEU(SOC). (JP 1-02)

Marine Logistics Command—The US Marines may employ the concept of the Marine Logistics Command (MLC) in major regional contingencies to provide operational logistic support, which will include arrival and assembly operations. The combat service support operations center will be the MLC's primary combat service support coordination center for units undergoing arrival and assembly. Also called MLC. (JP 1-02)

maritime pre-positioning force—A task organization of units under one commander formed for the purpose of introducing a MAGTF and its associated equipment and supplies into a secure area. The maritime pre-positioning force is composed of a command element, a maritime pre-positioning ships squadron, a MAGTF, and a Navy support element. Also called MPF. (MCRP 5-12C)

maritime pre-positioning force operation—A rapid deployment and assembly of a Marine expeditionary force in a secure area using a combination of strategic airlift and forward-deployed maritime pre-positioning ships. (JP 1-02)

maritime pre-positioning ships—Civilian-crewed, Military Sealift Command-chartered ships that are organized into three squadrons and are usually forward-deployed. These ships are loaded with pre-positioned equipment and 30 days of supplies to support three Marine expeditionary brigades. Also called MPS. (JP 1-02)

maritime special purpose force—A task-organized force formed from elements of a Marine expeditionary unit (special operations capable) and naval special warfare forces that can be quickly tailored to a specific mission. The maritime special purpose force can execute on short notice a wide variety of missions in a supporting, supported, or unilateral role. It focuses on operations in a maritime environment and is capable of operations in conjunction with or in support of special operations forces. The maritime special purpose force is integral to and directly relies upon the Marine expeditionary unit (special operations capable) for all combat and combat service support. Also called MSPF. (JP 1-02)

military operations other than war—Operations that encompass the use of military capabilities across the range of military operations short of war. These military actions can be applied to complement any combination of the other instruments of national power and occur before, during, and after war. Also called MOOTW. (JP 1-02)

mission—1. The task, together with the purpose, that clearly indicates the action to be taken and the reason therefore. 2. In common usage, especially when applied to lower military units, a duty assigned to an individual or unit; a task. (JP 1-02)

mission statement—A short paragraph or sentence describing the task and purpose that clearly indicate the action to be taken and the reason therefore. It usually contains the elements of who, what, when, and where, and the reason therefore, but seldom specifies how. (MCRP 5-12A)

mission type order—1. Order issued to a lower unit that includes the accomplishment of the total mission assigned to the higher headquarters. 2. Order to a unit to perform a mission without specifying how it is to be accomplished. (JP 1-02)

mobile defense—Defense of an area or position in which maneuver is used with organization of fire and utilization of terrain to seize the initiative from the enemy. (JP 1-02)

National Command Authorities—The President and the Secretary of Defense or their duly deputized alternates or successors. Also called NCA. (JP 1-02)

noncombatant evacuation operations—Operations directed by the Department of State, the Department of Defense, or other appropriate authority whereby noncombatants are evacuated from foreign countries when their lives are endangered by war, civil unrest, or natural disaster to safe havens or to the United States. Also called NEO. (JP 1-02)

obstacle—Any obstruction designed or employed to disrupt, fix, turn, or block the movement of an opposing force, and to impose additional losses in personnel, time, and equipment on the opposing force. Obstacles can be natural, manmade, or a combination of both. (JP 1-02)

operational control—Transferable command authority that may be exercised by commanders at any echelon at or below the level of combatant command. Operational control is inherent in combatant command (command authority). Operational control may be delegated and is the authority to perform those functions of command over subordinate forces involving organizing and employing commands and forces, assigning tasks, designating objectives, and giving authoritative direction necessary to accomplish the mission. Operational control includes authoritative direction over all aspects of military operations and joint training necessary to accomplish missions assigned to the command. Operational control should be exercised through the commanders of subordinate organizations. Normally this authority is exercised through subordinate joint force commanders and Service and/or functional component commanders. Operational control normally provides full authority to organize commands and forces and to employ those forces as the commander in operational control considers necessary to accomplish assigned missions. Operational control does not, in and of itself, include authoritative direction for logistics or matters of administration, discipline, internal organization, or unit training. Also called OPCON. (JP 1-02)

operational reach—The distance and duration across which a unit can successfully employ military capabilities. (JP 1-02)

operations security—A process of identifying critical information and subsequently analyzing friendly actions attendant to military operations and other activities to: a. identify those actions that can be observed by adversary intelligence systems. b. determine indicators hostile intelligence systems might obtain that could be interpreted or pieced together to derive critical information in time to be useful to adversaries; and c. select and execute measures that eliminate or reduce to an acceptable level the vulnerabilities of friendly actions to adversary exploitation. Also called OPSEC. (JP 1-02)

peace building—Post-conflict actions, predominately diplomatic and economic, that strengthen and rebuild governmental infrastructure and institutions in order to avoid a relapse into conflict. (JP 1-02)

peace enforcement—Application of military force, or the threat of its use, normally pursuant to international authorization, to compel compliance with resolutions or sanctions designed to maintain or restore peace and order. (JP 1-02)

peacekeeping—Military operations undertaken with the consent of all major parties to a dispute, designed to monitor and facilitate implementation of an agreement (cease fire, truce, or other such agreement) and support diplomatic efforts to reach a long-term political settlement. (JP 1-02)

peacemaking—The process of diplomacy, mediation, negotiation, or other forms of peaceful settlements that arranges an end to a dispute, and resolves issues that led to it. (JP 1-02)

position defense—The type of defense in which the bulk of the defending force is disposed in selected tactical localities where the decisive battle is to be fought. Principal reliance is placed on the ability of the forces in the defended localities to maintain their positions and to control the terrain between them. The reserve is used to add depth, to block, or restore the battle position by counterattack. (JP 1-02)

power projection—The application of measured, precise offensive military force at a chosen time and place, using maneuver and combined arms against enemy forces. (MCRP 5-12C)

rear operations—Military actions conducted to support and permit force sustainment and to provide security for such actions. (MCRP 5-12C)

reconstitution—Those actions that commanders plan and implement to restore units to a desired level of combat effectiveness commensurate with mission requirements and available resources. Reconstitution operations include regeneration and reorganization. (MCRP 5-12A)

regimental landing team—A task organization for landing comprised of an infantry regiment reinforced by those elements that are required for initiation of its combat function ashore. Also called RLT. (JP 1-02)

reserve—1. Portion of a body of troops that is kept to the rear, or withheld from action at the beginning of an engagement, in order to be available for a decisive movement. 2. Members of the Military Services who are not in active service but who are subject to call to active duty. (JP 1-02)

retirement—An operation in which a force out of contact moves away from the enemy. (JP 1-02)

retrograde movement—Any movement of a command to the rear, or away from the enemy. It may be forced by the enemy or may be made voluntarily. Such movements may be classified as withdrawal, retirement, or delaying action. (JP 1-02)

rules of engagement—Directives issued by competent military authority that delineate the circumstances and limitations under which United States forces will initiate and /or continue combat engagement with other forces encountered. Also called ROE. (JP 1-02)

screen—4. A security element whose primary task is to observe, identify and report information, and which only fights in self-protection. (JP 1-02)

sea control operations—The employment of naval forces, supported by land and air forces as appropriate, in order to achieve military objectives in vital sea areas. Such operations include destruction of enemy naval forces, suppression of enemy sea commerce, protection of vital sea lanes, and establishment of local military superiority in areas of naval operations. (JP 1-02)

sector—1. An area designated by boundaries within which a unit operates, and for which it is responsible. (JP 1-02)

security force—The detachment deployed between the main body and the enemy (to the front, flanks, or rear of the main body) tasked with the protection of the main body. The security force may be assigned a screening, guard, or covering mission. (MCRP 5-12C)

Service component command—A command consisting of the Service component commander and all those Service forces, such as individuals, units, detachments, organizations, and installations under that command, including the support forces that have been assigned to a combatant command or further assigned to a subordinate unified command or joint task force. (JP 1-02)

shaping—The use of lethal and nonlethal activities to influence events in a manner which changes the general condition of war to an advantage. (MCRP 5-12C)

special purpose MAGTF—A Marine air-ground task force organized, trained, and equipped with narrowly focused capabilities. It is designed to accomplish a specific mission, often of limited scope and duration. It may be any size, but normally it is a relatively small force - the size of a Marine expeditionary unit or smaller. Also called SPMAGTF. (JP 1-02)

spoiling attack—A tactical maneuver employed to seriously impair a hostile attack while the enemy is in the process of forming or assembling for an attack.

Usually employed by armored units in defense by an attack on enemy assembly positions in front of a main line of resistance or battle position. (JP 1-02)

strong point—A key point in a defensive position, usually strongly fortified and heavily armed with automatic weapons, around which other positions are grouped for its protection. (JP 1-02)

supporting effort—Designated subordinate unit(s) whose mission is designed to directly contribute to the success of the main effort. (MCRP 5-12C)

sustained operations ashore—The employment of Marine Corps forces on land for an extended duration. It can occur with or without sustainment from the sea. Also called SOA. (MCRP 5-12C)

strategic mobility—The capability to deploy and sustain military forces worldwide in support of national strategy. (JP 1-02)

support—1. The action of a force that aids, protects, complements, or sustains another force in accordance with a directive requiring such action. (JP 1-02)

synchronization—1. The arrangement of military actions in time, space, and purpose to produce maximum relative combat power at a decisive place and time. (JP 1-02)

tactical control—Command authority over assigned or attached forces or commands, or military capability or forces made available for tasking, that is limited to the detailed and, usually, local direction and control of movements or maneuvers necessary to accomplish missions or tasks assigned. Tactical control is inherent in operational control. Tactical control may be delegated to, and exercised at any level at or below the level of combatant command. Also called TACON. (JP 1-02)

tactical recovery of aircraft and personnel—A mission performed by an assigned and briefed aircrew for the specific purpose of the recovery of personnel, equipment, and/or aircraft when the tactical situation precludes search and rescue assets from responding and when survivors and their location have been confirmed. Also called TRAP. (MCRP 5-12C)

tempo—The relative speed and rhythm of military operations over time. (MCRP 5-12C)

warfighting functions—The six mutually supporting military activities integrated in the conduct of all military operations are:1. command and control-The means by which a commander recognizes what needs to be done and sees to it that appropriate actions are taken. 2. maneuver-The movement of forces for the purpose of gaining an advantage over the enemy. 3. fires-Those means used to delay, disrupt, degrade, or destroy enemy capabilities, forces, or facilities as well as affect the enemy's will to fight. 4. intelligence-Knowledge about the enemy or the surrounding environment needed to support decisionmaking. 5. logistics-All activities required to move and sustain military forces. 6. force protection-Actions or efforts used to safeguard own centers of gravity while protecting, concealing, reducing, or eliminating friendly critical vulnerabilities. Also called WF. (MCRP 5-12C)

withdrawal operation—A planned retrograde operation in which a force in contact disengages from an enemy force and moves in a direction away from the enemy. (JP 1-02)

Bibliography

Fitton, Robert A., (ed.), *Leadership: Quotations from the Military Tradition*, Westview Press, Boulder, CO, 1990.

Heinl, Robert Debs, Jr., Col., USMC, Retired, *Dictionary of Military and Naval Quotations*, United States Naval Institute, Annapolis, MD, 1966.

Huntington, Samuel P., *National Policy and the Transoceanic Navy, U.S. Naval Institute Proceedings*, Menasha, WI, 1954.

Stokesbury, James L., *A Short History of the Korean War*, William Morrow and Company, Inc., New York, 1988.

United States. House. 10 US Code, *Armed Forces*, Washington, DC, GPO.

10 US Code, Subtitle A, *Goldwater-Nichols Department of Defense Reorganization Act of 1986*.

18 US Code, *Posse Comitatus Act*.

— Office of the Secretary of Defense, Washington, DC, GPO.

Functions of the Department of Defense and Its Major Components, DODD 5100.1, 1987.

Forces for Unified Commands, Secretary of Defense Memorandum.

— Office of the Joint Chiefs of Staff, Washington, DC. GPO.

Command and Control for Joint Air Operations, JP 3-56.1, 1994.

Department of Defense Dictionary of Military and Associated Terms, JP 1-02, 2001.

Doctrine for Joint Operations, JP 3-0, 1995.

Doctrine for Logistic Support of Joint Operations, JP 4-0, 2000.

Joint Doctrine for Amphibious Operations, JP 3-02, 1992.

Joint Doctrine for Military Operations Other Than War, JP 3-07, 1995.

Joint Doctrine for Rear Area Operations, JP 3-10, 1996.

Joint Vision 2020, 2000.

Unified Action Armed Forces (UNAFF), JP 0-2, 2000.

— United States Marine Corps. Washington, DC, GPO.

Building a Corps for the 21st Century, Concepts and Issues '98, 1998.

Campaigning, MCDP 1-2, 1997.

Command and Control, MCDP 6, 1996.

Communications and Information Systems, MCWP 3-40.3, 1998.

Componency, MCDP 1-0.1, 1998.

Expeditionary Operations, MCDP 3, 1998.

Joint Force Land Component Commanders Handbook, MCWP 3-40.7, 2001.

Intelligence, MCDP 2, 1997.

Logistics, MCDP 4, 1997.

Logistics Operations, MCWP 4-1, 1999.

Marine Corps Manual, 1980.

Marine Corps Planning Process, MCWP 5-1, 1998.

Marine Corps Strategy 21, 2000.

Planning, MCDP 5, 1997.

Policy for Marine Expeditionary Unit (Special Operations Capable), MCO 3120.9A, 1997.

Rear Area Operations, MCWP 3-41.1, 2000.

Small Wars Manual, NAVMC 2890, 1967.

Tactical Level Logistics, MCWP 4-11, 2000.

Tactics, MCDP 1-3, 1997.

Warfighting, MCDP 1, 1997.

Vandegrift, Alexander A., *The Marine Corps in 1948, United States Naval Institute Proceedings*, Menasha, WI, 1948.

www.ingramcontent.com/pod-product-compliance
Lightning Source LLC
Chambersburg PA
CBHW080531170426
43195CB00016B/2526